TERRARVM
Tartaria
ASIA
Mongol
Cathaio
China
Turcheſtan
Perſia
India orientalis
Arabia
EVROPA
AFRICA
Nubia
MAR DI INDIA
IS NONDVM COGNITA.
VS HVMANIS, CVI AETERNITAS
SIT MAGNITVDO. CICERO:

Michael Mann / Ineke Phaf-Rheinberger (eds)

Beyond the Line

Cultural Narratives of the Southern Oceans

Beyond the Line

Cultural Narratives of the Southern Oceans

edited by
Michael Mann and
Ineke Phaf-Rheinberger

Neofelis Verlag

German National Library Cataloguing in Publication Data
A catalogue record for this book is available from the German National Library:
http://dnb.d-nb.de

Cover Design: Marija Skara
Endpapers: Abraham Ortelius: *Typus Orbis Terrarum,* 1572.
Printed in EU
ISBN: 978-3-943414-14-1

Contents

II. Narrating the Ocean

Introduction: 'Beyond the Line' – Constructions and Concepts with Focus on the Southern Oceans

Michael Mann / Ineke Phaf-Rheinberger

1. Beyond which Line?

The title of this volume, "Beyond the Line", refers to the fact that all its contributions address constructions and concepts elaborated with regard to the Southern Oceans. "Beyond the Line" draws on the phrase "no peace beyond the line" that appeared in the peace treaty of Cateau-Cambrésis which was concluded in 1559. Both phrases became quite well known in historical research. However, the question remains what line is actually referred to? Many essays have been written on this subject and James A. Williamson, in his famous study *Hawkins of Plymouth*, seems to provide an ultimate solution:

> In peace or war in Europe there was no peace beyond the line. The phrase is often quoted by people who do not explain what line they mean. The Tropic of Cancer will not by itself answer the question, neither will the lines of demarcation. "Line" is in fact a misquotation, which should be "lines". The "lines of amity" were verbally agreed upon by the French and Spanish negotiators of the Peace of Cateau-Cambrésis in 1559. They were to be the Tropic of Cancer and the prime meridian passing through Ferro in the Canaries. On the European side of both lines the treaty was to be binding; west and south of them it was to be disregarded. The agreement was a belated recognition of what had long been the practice.1

The peace treaty of Cateau-Cambrésis did not mention any concrete line as a line of demarcation. Only an oral agreement added to the treaty is said to have indicated some lines of amity. Academic research

1 James A.Williamson: *Hawkins of Plymouth. A New History of Sir John Hawkins and of the Other Members of his Family Prominent in Tudor England.* 2nd ed. London: Black 1969, p. 46.

maintained, first, that the addition to the peace treaty did in fact refer to two lines fixing an imaginary 'Atlantic cross' based on the Tropic of Cancer and the European understanding of a global Meridian and, second, that peace was only valid on the European side of the lines.[2] Most legal historians and maritime historians followed this interpretation rather uncritically. More recently it is argued that hardly any concrete juridical and legal idea existed about the line.[3]

Williamson resumes that from the sixteenth century on the phrase "no peace beyond the line" referred to European sovereigns' well-established practice of issuing "letters of marque and reprisal". It was also in some way taken for granted that such letters created and demarcated the space for buccaneering, freebooting, and piracy in a maritime space defined by meridian and longitude. The line was constructed on the Atlantic Ocean and made the maritime territories a legal, political, and cultural construction in cartography since Early Modern History. European peace treaties among sovereigns were in almost all cases written documents and the stipulation of the validity of peace and war overseas was not congruent with contemporary European 'international' law. This might explain why no preserving text of the oral agreement exists. In essence, however, European bilateral and multilateral peace treaties were considered to be of universal value. But in reference to trade and commercial matters it was different. The Spanish persistently defended de jure regulations against de facto practice and implementation to ensure their trading monopoly with the colonies in the Caribbean and Central America, thereby trying to defend a space for their own mercantile operations.[4]

"The line" only became a matter of bilateral and multilateral regulations in the course of the seventeenth century. After the defeat of the Spanish Armada in 1588 and the founding of the Northern European East India trading companies such as the English East India Company in 1600 and the Dutch Vereenigde Oostindische Compagnie in 1602,

2 Frances Gardiner Davenport: *European Treaties Bearing on the History of the United States and its Dependencies*, 4 vols. Washington, DC: Carnegie Institute 1917–1937, vol. 1: (1716–1815), p. 120.

3 Jörg Fisch: *Die europäische Expansion und das Völkerrecht. Die Auseinandersetzungen um den Status der überseeischen Gebiete vom 15. Jahrhundert bis zur Gegenwart.* Stuttgart: Steiner 1984, p. 59.

4 Fisch: *Die europäische Expansion*, pp. 74–77, 82–86.

particularly the Dutch claimed free access to all oceans of the globe. This debate basically continued the old legal question of *mare clausum* and *mare liberum*.[5] The Dutch as well as the English claimed the right to free and peaceful trade on the seas of the world, which did not imply the establishment of peaceful conditions in overseas territories. Hugo Grotius became famous for his treatise "Mare Liberum" published in 1609 in which he rejected the Papal Treaty of Tordesillas (1494) that had divided the globe into a Spanish and a Portuguese sphere of expansion and occupation. In contrast, Grotius demanded open access for all seafaring nations to the oceans of the globe.[6] Interestingly, at the same time, the line (always singular, never plural "lines") began to be increasingly mentioned in European treaties. It was understood that the Tropic of Cancer, i. e., the equinoctial, marked off all colonial territories, roughly distinguishing between Europe and the rest of the world.

A second line was not explicitly addressed but implicitly thought of: the meridian passing through the Canary Islands and the Azores. According to present day geography, it is not possible to draw the line that way but, in accordance with nautical knowledge in the sixteenth century when methods of determining latitudes did not exist,[7] cartographers placed the Azores on the same meridian as the Canary Islands.[8] In a restricted sense, as it was understood at that time, "beyond the line" basically confined the broad maritime space of the Caribbean,[9] which became a legendary region for European adventurers seeking their fortune. This might explain the enormous success of the first fiction book on pirates, *De Americaensche Zee-Rovers*, written by the French Huguenot Alexandre Exquemelin and published in Amsterdam in

5 Garrett Mattingly: No Peace Beyond What Line? In: *Transactions of the Royal Historical Society* 13 (1963), pp. 145–162, cf. pp. 145–147.

6 In 1604 Hugo Grotius started writing his treatise on *De Jure Praedae* (On the Right of Prize) which was only published in 1868. The chapter on "Mare Liberum" however was published in 1609 and became famous as well as influential with respect to the emerging international law. Cf. Hugo Grotius: *The Freedom of the Sea – Or, The Right Which Belongs to the Dutch to Take Part in the East Indian Trade.* New York: Oxford University Press 1916.

7 David Sobel: *Longitude. The True Story of a Lone Genius Who Solved the Greatest Scientific Problem of His Time.* Harmondsworth etc.: Penguin 1995.

8 Mattingly: No Peace Beyond What Line?, p. 148.

9 Fisch: *Die europäische Expansion*, pp. 119–122.

1678. The book was immediately translated into German, Spanish, English, and French and has been often rewritten ever since.[10] While it seems clear that in Early Modern History "the line" was the Tropic of Cancer, the northern equinoctial and the equator were, however, often mixed up, both denominations being used at the same time in different documents.

During the eighteenth century the equator became "the line", indicating the growing maritime dominance of England, a country that at all refused to accept the Caribbean as a maritime space outside the sphere of treaties.[11] It was also taken for granted that "beyond the line" implied a maritime space with 'internationally' valid rules and regulations governing shipping, trade, and commerce, which included the mutual acceptance of trading monopolies. "Beyond the line" was never a legally fixed category, but a political instrument to enforce favourable conditions when negotiating treaties dealing with overseas maritime and territorial spaces.[12] At the same time, the meridian disappeared as the complementary line of the 'Atlantic cross' (the Tropic of Cancer and the meridian). From the eighteenth century on, the equator separated a northern civilised "West" from the sharply contrasting rest of the world, in particular the "East". Along with this change, the meaning of the line shifted from a geopolitical to a more cultural one.

As part of the cultural divide, rituals on board ships marked the transition into another world during the late seventeenth and the eighteenth centuries. According to Francis Rogers, when sailing to the "West Indies"—and therefore crossing the Tropic of Cancer!—at the beginning of the eighteenth century, all people on board the ship had the choice of either paying a bottle of brandy and a pound of sugar or, instead, of being ducked into the ocean. The person who refused to pay was plunged into the water from the main yardarm secured only by a rope and was then hauled up again. This procedure was repeated three times.[13] And Jean François Michel described another

10 James A. Arnold: From Piracy to Policy: Exquemelin's Buccaneers and Imperial Competition in America. In: *Review* 40,1 (2007), pp. 9–20.

11 Mattingly: No Peace Beyond What Line?, pp. 154–156.

12 Fisch: *Die europäische Expansion*, pp. 113–119, 131.

13 "The Journal of Francis Rogers". In: Bruce S. Ingram (ed.): *Three Sea Journals of Stuart Times*. London: Constable 1936, pp. 143–230, here p. 152. The edited volume includes the diary of Dawtrey Cooper, 1628, kept during the expedition to relieve La

ritual that took place on the *Burg von Emden* when it crossed the equator some decades later. The captain ordered a bowl of seawater and a plate with some zwieback and salt to be placed on a table in the back. The gunman was asked to prepare a gunshot to greet the arriving Neptune. Neptune, obviously a member of the crew, came down from the foresail mast, covering his head with a sheep's coat topped by a crown made of paper. His face, hands, and feet were blackened, his beard consisted of old ropes and he carried a trident. The captain invited Neptune to the table. The crew, which had assembled on deck, was asked to sign a paper pledging a donation for the poor of the town of Emden (situated in northern Germany). Everyone signed and was offered a piece of zwieback and salt by the first officer. The captain then 'baptised' everyone with a handful of water, in some instances with a ladle of water.[14] At some stage at the end of the eighteenth century, these rituals disappeared, at least from the travelogues.

Walter Scott was the first author to argue in a leaflet on Walter Raleigh in 1806 that it was generally assumed that in Raleigh's time "there was no peace beyond the line".[15] From then on the phrase became part of an English invented historical and cultural tradition romanticising the pirates of the Caribbean and the 'Age of Sail' on the one hand and, on the other hand, transforming them into national heroes by turning Her Majesty's agent Sir Francis Drake into the first and most prominent English opponent to Spanish colonial claims in America. Thus, during the 'Age of Steam', the phrase "no peace beyond the line" became prominent as a catchphrase, at least in the English cultural and literary tradition.[16]

In short, it is quite interesting to note that initially "the line" once indicated a political, juridical or military category defining a maritime space of commercial and mercantile law understood or claimed as

Rochelle; the journals of Jeremy Roch, 1666, describing voyages and adventures at sea under Charles II, James II and William III; and the diary of Francis Rogers, kept on his voyages to the West Indies, the East Indies, and elsewhere in 1703 and 1704.

14 Journal du voyage a la Chine, et courte description de la ville d'Embden par J. F. Michel (unpubl. manuscript), parts published and translated into German, *vide* Dok. 35: Der Unterkaufmann Jean François Michel über die Äquatortaufe auf dem Chinafahrer *Burg von Emden* (1752). In: Eberhard Schmitt (ed.): *Indienfahrer 2. Seeleute und Leben an Bord im Ersten Kolonialzeitalter*. Wiesbaden: Harrassowitz 2008, pp. 240–245.

15 Fisch: *Die europäische Expansion*, pp. 141–145.

16 Mattingly: No Peace Beyond What Line?, pp. 161–162.

mare liberum. However, this claim never corresponded with reality. In the twentieth century "the line" definitely refers to the equator and had become a legal and, additionally, racial and civilisational category defining the "West's" legal differences with respect to labour and civic rights. Today nobody speaks of the "line" and many present-day contemporaries have forgotten its existence as well as the meaning of the phrase. But "the line" is still there though not explicitly referred to. For example—and one example may suffice at this point—when Western media or politicians of the G8 speak of the "Global South", they roughly denominate what were termed "developing countries" during the second half of the twentieth century, all of which were former colonies or countries indirectly ruled by European or imperial(ist) powers. And of course, the "Global South" is a zone where things are entirely different.

2. The "Global South" in the Twentieth-First Century

At the end of the twentieth century, the changing role of the "Global South" ran parallel with geopolitical modifications in which the ocean as such became an increasingly important space. This was not the same in the past. Research projects usually chose as its objects a country's people, landscape, and societies. Until recently, seas and oceans were considered only to the degree that they offered communication routes in the context of economic relations, i. e., trade routes and commercial connections. In historical terms, seas, like mountain ranges, tended to be viewed more as separating than as connecting elements. Yet, as many recent studies have demonstrated, seas and oceans were zones of cultural encounter and communication. This was first elaborated for the Mediterranean Sea[17] and, meanwhile, an increasing number of studies of this kind on the Atlantic[18] and Indian Oceans[19] have

17 Fernand Braudel: *Das Mittelmeer und die mediterrane Welt in der Epoche Philipps II*, 3 vols. Frankfurt am Main: Suhrkamp 1990 [*La Méditerranée et le monde méditerranéen à l'époque Philippe II*. Paris: 1949].

18 Douglas R. Egerton / Alison Games / Jane G. Landers / Kris Lane / Donal R. Wright (eds): *The Atlantic World: A History*. Wheeling, IL: Harlan Davidson 2007; John H. Elliott: *Empires of the Atlantic World; Britain and Spain in America 1492–1830*. Yale, NH: Yale University Press 2009; Benjamin, Thomas: *The Atlantic World. Europeans, Africans, Indians and Their Shared History*, 1400–1900. Cambridge: Cambridge University Press 2009; Leos Müller / Philipp Robinson Rössner / Toshiaki Tamaki (eds): *The Rise of the Atlantic Economy and the North Sea-Baltic Trade, 1500–1800*. Stuttgart: Steiner 2011.

19 Kirti N. Chaudhuri: *Asia Before Europe. Economy and Civilisation of the Indian Ocean from the Rise of Islam to 1750*. Cambridge: Cambridge University Press 1990; Kenneth

come out. At present, sea-borne migrations are even considered very relevant for having established networks on a global scale thus constituting a fundamental regularity in the history of mankind.[20]

A major historiographical event was the presentation of a special issue of *The Geographical Review*,[21] in which the editors Martin W. Lewis and Karen Wigen offer a "maritime response to the crisis in area studies". Area studies had come under pressure from the emerging field of global studies that emphasies that all the regions of the world should play a role, restoring those that had been invisible historically on the conventional maps. Concentrating on the switches of focus in research, Lewis and Wigen claim that geography in particular was absent in global studies since the 1990s and propose correcting this by taking the maritime dimension into account. In his essay "Dividing the Ocean Sea", Lewis examines the history of global hydrography from the classical tradition until today concluding that "the Western conceptualisation of the oceans may better be described as one of aimless wanderings. In this Foucauldian story line, different ways of dividing and labeling the sea come in and out of fashion, each successive view reflecting the epistemic environment of its time without adding any cumulative conceptual purchase".[22]

At the Center Modern Orient (ZMO) in Berlin, a discussion took place at the beginning of this millennium on reframing area studies around an ocean basin, specifically the Indian Ocean. The volume *Translocality. The Study of Globalising Processes from a Southern Perspective* presents some results of this research.[23] The papers accuse the discussion of globalisation of having a Eurocentric world-view and of a renewed

McPherson: *The Indian Ocean. A History of People and the Sea.* Delhi et al.: Oxford University Press 1998; Michael N. Pearson: *The Indian Ocean.* London / New York: Routledge 2003; Dietmar Rothermund / Susanne Weigelin-Schwiedrzik (eds): *Der Indische Ozean. Das afro-asiatische Mittelmeer als Kultur- und Wirtschaftsraum.* Wien: Mandelbaum 2004; Sugata Bose: *A Hundred Horizons. The Indian Ocean in the Age of Global Empire.* Cambridge, MA / London. Harvard University Press 2006.

20 Donna R. Gabaccia / Dirk Hoerder (eds): *Connecting Seas and Connected Ocean Rims. Indian, Atlantic, and Pacific Oceans and China Seas Migrations from the 1830s to the 1930s.* Leiden: Brill 2011.

21 Martin W. Lewis / Karen Wigen: A Maritime Response to the Crisis in Area Studies. In: *The Geographical Review* 89,2 (1999), pp. 161–168.

22 Lewis: Dividing the Ocean Sea. In: Ibid, pp 188–214, here p. 210.

23 Ulrike Freitag / Achim von Oppen (eds): *Translocality. The Study of Globalising Processes from a Southern Perspective.* Leiden: Brill 2010.

appropriation of the 'postcolonial' present. The latter is partly due to the fact that studies on Asia, Africa, and Latin America have seldom focused on these regions' relations among each other. This is why the editors suggest the term "translocality", so that, from the beginning, these areas in different continents are not grasped as separate from each other, but as regions with societies that are connected by means of "histoires croisées" or "entangled histories".

Today, for such entangled connections, which can be effective even across oceans, terms like "marginal mobilities", "spaces on the move", and "alternative globalities" are used in reference to the circulation of labour forces (slaves, coolies), the phenomenon of piracy, travelers' reports, living spaces on ships, and the transport of goods. The history of transport has a particular complexity in countries with a colonial background. Ravi Ahuja discusses some of its problematic aspects in *Pathways of Empire. Circulation, 'Public Works' and Social Space in Colonial Orissa, c. 1780-1914.*[24] He claims a critical reorientation that looks at the "historical constellations of social forces behind supposedly timeless technical necessity" and discerning a "plurality of distinct and potentially conflicting appropriations behind the appearance of consensual utilisation".[25] Ahuja emphasises the production of "social space", a concept he borrows from Henri Lefebvre,[26] to trace the changes and repetitive socio-spatial practices in long-term processes of circulation. The history of transport in India was one-sidedly concerned with the production of railways and canals, but had neglected "local" roads and sea spaces, among other things. In accordance with Ahuja's argument, the ocean, too, might serve as a conceptualised space: "the space of scientists, planners, urbanists, technocratic subdividers and social engineers, as of a certain type of artist with a scientific bent".[27]

Without a doubt, oceans that wash the coasts of continents have always been transportation routes. Arab, African, and Asian ships navigated the Indian Ocean for more than a millennium and, from 1500 on,

24 Ravi Ahuja: *Pathways of Empire. Circulation, "Public Works" and Social Space in Colonial Orissa, c. 1780–1914*. Hyderabad: Orient Blackswan 2009.

25 Ibid., pp. 3–4.

26 Henri Lefebvre: *The Production of Space*, trans. by Donald Nicholson-Smith. Oxford / Cambridge, MA: Basil Blackwell 1991 [*La production de l'espace*. Paris: 1974].

27 Ahuja: *Pathways of Empire*, p. 113.

were joined by European trading ships.[28] In the course of the consolidation of the dynasties in the Ottoman, Safavid, and Mogul Empires and of the Iberians' expansion into the Atlantic and Indian Oceans since the beginning of the sixteenth century, the Atlantic as well as the Indian Ocean became spaces occupied by the growing number of trading ships. Increasing interaction led to quantitatively and, even more important, qualitatively denser networks that were not based solely on economic ties and that to a great degree developed out of pre-colonial structures.[29]

Hence, in these emerging imaginative and / or real networks, sub-Saharan, southern Africa has hardly ever been perceived as an active (and attractive) participant and seems to have been, for centuries, only a supplier of millions of slaves to be kidnapped and shipped to the Americas.[30] A recent overall study shows that there was a well-organised slave trade in the Indian Ocean as well.[31] It should be noted in this context that the slave trade indeed played an important economic, social, and not least demographic role in some regions of Africa. Notwithstanding, along with the slave trade, especially in the Indian Ocean, there were much more diverse forms of interregional and intercontinental economic relations. In general, the macro-economic exchange always also contributed to cultural transfers and intertwining. Slave trafficking is not just about the transport of a labour force from one place to another, but also implies the transfer of cult, culture, and creativity as well.

Because of the historical narrative of slavery and of the disease-infested 'black continent' that was veritably carved up between the European imperial powers in the notorious "Scramble for Africa" at

28 Compare Kenneth Pomeranz / Steven Topik: *The World that Trade Created. Society, Culture, and the World Economy.* New York / London: Shape 2006. Jürgen Osterhammel / Niels P. Petersson: *Geschichte der Globalisierung. Dimensionen, Prozesse, Epochen.* München: C. H. Beck 2003.

29 Roderich Ptak: *Die maritime Seidenstraße. Küstenräume, Seefahrt und Handel in vorkolonialer Zeit.* Munich: C. H. Beck 2007; Diogo Ramada Curto / Francisco Bethencourt (eds): *Portuguese Oceanic Expansion, 1400–1800.* Cambridge: Cambridge University Press 2007.

30 Herbert S. Klein: *The Atlantic Slave Trade.* Cambridge: Cambridge University Press 2002.

31 Michael Mann: *Sahibs, Sklaven und Soldaten. Geschichte des Menschenhandels rund um den Indischen Ozean.* Darmstadt: WBG 2011.

the end of the nineteenth century, essential parts of the history of Africa seem to have been lost or subsumed in the colonial-imperial narrative. The world's basically negative image of Africa is only gradually being revised after the founding of independent republics in large parts of the continent.[32] African states have their own representatives in the (post-colonial) bodies of world organisations and the demand for their raw materials and energy sources is ever growing.[33] Despite the huge economic and in particular social problems that some states and their governments have to battle with, this old role has taken on new dimensions creating new forms of self-understanding and self-respect as well.

Along with this changing role of Africa as a global player, one can observe an increasing sensitivity to the historical dimension of the seas and of water in general for human history. Although Lewis and Wigen regarded the "maritime approach" as only one of the possible alternatives to area studies, in the meantime, this approach has become ever more relevant. Controlling terrestrial water seems to be a local, bilateral and international problem; the same seems true for estuaries, coastal waters, and the high seas. The contributions in the three-volume series *A History of Water* investigate historical and contemporary dealings with the organisation of water in human societies.[34] These studies can serve as spurs to study the cultural history of the oceans, for which the term "invisible continent" has been created.[35] In relation to "water management around the globe" and its "water narratives",[36] a changed environmental consciousness has emerged recently. Headlines like "Forscher schlagen Alarm. Weltmeere steuern auf Kollaps

32 Leonardo Villalón / Peter Von Doepp (eds): *The Fate of Africa's Democratic Experiments. Elites and Institutions.* Bloomington: Indiana University Press 2005; Martin Meredith: *The State of Africa. A History of Fifty Years of Independence.* The Free Press 2005.

33 John H. Ghazvinian: *Untapped. The Scramble for Africa's Oil.* Orlando: Harcourt 2007; Duncan Clarke: *Crude Continent: The Struggle for Africa's Oil Prize.* London: Profile Books 2009. African Development Bank / African Union (eds): *Oil and Gas in Africa.* Oxford / New York: Oxford University Press 2009.

34 Terje Tvedt et al. (eds): *A History of Water*, 3 vols. London: Tauris 2006–2011.

35 Marc Muguet: *Après Cousteau. Le futur de l'Océan.* Paris: Éditions Yago 2010.

36 Terje Tvedt: Introduction: Water History is World History. In: Id. / Eva Jakobssen (eds): *A History of Water*, vol. 1: Water Control and River Biographies. London: Tauris 2006, pp. IX–XXIII.

zu"[37] (Researchers sound the alarm. Oceans heading for collapse) are no rarity. Pollution by waste discharge, overfishing, garbage disposal, container ship disasters and the rise of the sea level are prominent themes in this discourse about oceans, which Marc Muguet has called the "clef du future".[38] No one can foresee what the polluted water from the nuclear power plant in Fukushima will do to the Pacific Ocean. Genetic analysis of deep-sea giant squid[39] and the inauguration of an ocean park, the Great Barrier Reef, declared a Cultural World Heritage by the UNESCO, are new phenomena. In addition to that, we observe such contrastive panoramas of boat people in the Mediterranean, the Caribbean or the Indian Ocean trying to reach more prosperous coasts in dramatic circumstances whereas cruise ships attract more tourists than ever for relaxing during their holidays on these same waters.

3. Literature, Memory, and the Sea

Many studies start by contradicting the term "invisible continent" when addressing the oceanic world. This approach goes together with a broader projection toward the cultural history of the ocean. Rila Mukherjee, for instance, edited the volume *Oceans Connect: Reflections on Water Worlds Across Time and Space.*[40] Her own contribution has the title "Chasing the Many Faces of the Marine Goddess",[41] in which she analyses folkstales and architectural ornaments. In addition, the terminology of water worlds is also influencing the idiom for all types of communication media and routes, operating with words such as "airport", "airship", and "surfing" or "navigating" in the Internet. The construction of a virtual space, in which the oceanic world plays a role, came into being from the sixteenth century on, when an increasing visual and literary construction of these spaces can be observed

37 Weltmeere steuern auf Kollaps zu. http://www.stern.de/wissen/natur/forscher-schlagen-alarm-weltmeere-steuern-auf-kollaps-zu-1697922.html (accessed February 24, 2014).

38 Marc Muguet: *Après Cousteau*, pp. 293–295.

39 Katharina Dellai-Schöbi: Geheimnisvoller Riesenkalmar. Genanalysen aus der Tiefsee. http://www.nzz.ch/wissen/wissenschaft/genanalysen-aus-der-tiefsee-1.18065139 (accessed February 24, 2014).

40 Rila Mukherjee (ed.): *Oceans Connect: Reflections on Water Worlds Across Time and Space.* Delhi: Primus Books 2013.

41 Mukherjee: *Après Cousteau*, pp. 39–52.

going far beyond cartography and the first forms of European travel literature.[42]

Interestingly, in the visual art of the "Dutch Seaborne Empire",[43] seascapes as well as landscapes became autonomous painting genres and were quite successful on the market. This outstanding role of visual technology, the "Art of Describing,"[44] explains the output from the overseas expedition of Count Maurits von Nassau to Brazil, Latin America, and Africa in the service of the Dutch West India Company from 1637 to 1644, offering a treasure horde of illustrations, paintings, and meticulous descriptions of people and landscapes overseas. Fauna, flora, and nature were depicted, as were the different types of workers. We owe to this collection the first modern city views of Luanda and Recife, the depictions of churches, of houses, and of public places, and the portrayal of different grades of cultural diversity and affiliation.[45]

Today, seas and oceans are conceptualised as shaping entangled realities through the exchange of people, goods, thoughts, and ideas. This dynamic becomes a special dimension in literature and art. The sea as a 'living entity' and as an autonomous yet well connected zone of cultural exchange and its own culture was already described in eighteenth-century texts such as Elizabeth Post's epistolary novel in three volumes *Reinhart, of natuur en godsdienst* (1791–1792), the first 43 letters of which are written by Reinhart on the ship sailing from Holland to Guyana in South America. The author addresses fish as well as birds, and let Reinhart develop his philosophy on divine Creation:

42 Compare for India / Southern Asia Gita Dharampal-Frick: *Indien im Spiegel deutscher Quellen der frühen Neuzeit (1500–1750). Studien zu einer interkulturellen Konstellation.* Tübingen: Niemeyer 1994.

43 C. R. Boxer: *The Dutch Seaborne Empire 1600–1800.* London: Penguin 1990 [1965].

44 Svetlana Alpers: *The Art of Describing. Dutch Art in the Seventeenth Century.* London: Penguin 1989.

45 E. van den Boogaart / H. R. Hoetink / P. J. P. Whitehead (eds): *Johan Maurits van Nassau-Siegen 1604–1679. A Humanist Prince in Europe and Brazil.* The Hague: The Johan Maurits van Nassau Stichting 1979; Pedro Corrêa do Lago / Bia Corrêa do Lago: *Frans Post 1612–1680. Obra Completa.* Rio de Janeiro: Capivara Editora 2006; Rebecca Parker Brienen: *Visions of Savage Paradise. Albert Eckhout, Court Painter in Colonial Dutch Brazil.* Amsterdam: Amsterdam University Press 2006; Ineke Phaf-Rheinberger: *Vision* and the Representation of Africans – On Historical Encounters between Science and Art. In: *History of Philosophy and the Life Sciences* 35 (2013), pp. 53–61.

> The journey led through the Canal (Dover), southwards along Madeira, the Canary Islands, from there crossing the Atlantic Ocean in the direction of Guiana. Reinhart describes the sea under variable circumstances, at night and by day, windless or under the burning sun, with mist and storm. He also dedicates a long, partly speculative consideration to underwater life in the ocean. The sea always demonstrates God's greatness, his omnipotence and his goodness. When the ship is at risk, for example in misty weather or during a heavy tempest, Reinhart's fear disappears due to this faith in Providence: if God does not want it, no one can be hurt. Everything on the sea – from the rolling of the waves to the death of a seaman – reminds Reinhart of the vanity of man and creation.[46]

Without a doubt, the sea has been most intensively studied in English-speaking literature and in the tradition of Paul Gilroy's *Black Atlantic* in particular.[47] We find a syllabus of courses on "The Sea in Nineteenth-Century English and American Literature" in the Internet.[48] Many essays and much literature deal with the sea, with Melville's famous *Moby Dick* as the foremost example. Following up on this tradition in Anglophone literature, Bernard Klein and Gesa Mackenthun organised two volumes in which they examined the sea as a contact zone, a concept introduced by Mary Louise Pratt in her book *Imperial Eyes*, analysing its role in history.[49] Numerous authors concentrate on the Caribbean, the Pacific, or the Indian Ocean.[50] Amitav Ghosh

46 "De reis leidt achtereenvolgens door het Kanaal (Dover), zuidwaarts langs Madeira, de Canarische eilanden; vandaar wordt de Atlantische Oceaan overgestoken richting Guiana. Reinhart beschrijft de zee onder wisselende omstandigheden: bij nacht en dag, bij windstilte en brandende hitte, bij mist en bij storm. Ook aan het onderwaterleven in de oceaan wijdt hij een lange, deels speculatieve beschouwing. Altijd demonstreert de zee Gods grootheid, almacht en goedheid. Wanneer het schip risico's loopt, zoals bij mist of zware storm, wordt Reinharts angst weggenomen door zijn vertrouwen in de Voorzienigheid: buiten Gods wil kan er niemand iets overkomen. Van het voortrollen van de golven tot de dood van een matroos doet op zee Reinhart alles herinneren aan de vergankelijkheid van mens en schepping." (Bert A. N. Paasman: *Nederlandse Literatuur en slavernij ten tijde van de Verlichting*. Leiden: Nijhoff 1984, p. 13.)

47 Paul Gilroy: *The Black Atlantic. Modernity and Double-Consciousness*. Cambridge, MA: Harvard University Press 1993.

48 The Sea in Nineteenth-Century English and American Literature. http://www.enotes.com/topics/sea-nineteenth-century-english-and-american/critical-essays (accessed October 14, 2013).

49 Bernard Klein / Gesa Mackenthun (eds): *Das Meer als kulturelle Kontaktzone. Räume, Reisende, Repräsentationen*. Constance: ULV 2003; Bernard Klein / Gesa Mackenthun (eds): *Sea Changes. Historicizing the Ocean*. New York / London: Routledge 2004; Mary Louise Pratt: *Imperial Eyes. Travel Writing and Transculturation*. London / New York: Routledge 1992.

50 *Wasafiri* 66 (2011): Indian Oceans; Elizabeth DeLoughrey: *Routes and Roots: Navigating Caribbean and Pacific Island Life*. Honolulu: University of Hawaii Press 2007.

traverses the coasts of Southeast Asia in his novels *The Sea of Poppies* and the *River of Smoke*[51] on the ship named *Ibis*, a trilogy whose third volume has not come out yet. In Ghosh's books, the reconstruction of the nineteenth century on the sea powerfully demonstrates the fluid line between legal and illegal situations, the Clash of Civilisations and the Tower of Babel,[52] the intermediate zone in which people from all places come together with different goals in mind.

The Tower of Babel, of course, was a favorite topos in paintings in Early Modern History. This period created a series of new creole and pidgin languages, which are still spoken today and, in some cases, declared a national language, such as Kapverdianu in Cap Verde and Papiamentu in Curaçao. "The Burden of English"[53] becomes more diversified and, in the last decade, several studies address this oceanic connection beyond English-speaking literature, such as *The French Atlantic Triangle: Literature and Culture of the Slave Trade* (2008)[54] by Christopher Miller, which focuses on French literature since the Enlightenment. Miller's chapter on "African 'Silence'" is especially interesting, because he examines the "borderlines between utterance in certain works of Francophone African literature and film that are concerned with the slave trade".[55]

Other books, *The "Air of Liberty". Histories of the South Atlantic Past* and *Historias enredadas. Representaciones asimétiricas con vista al Atlántico*[56] discuss Dutch-, Spanish-, and Portuguese-language entanglements in America and Africa. The volume *Les empires de l'Atlantique (XIXe-XXIe siècles). Figures de l'autorité impériale dans les lettres d'expression européenne de l'espace atlantique*[57] comprises essays on the Dutch, English, and Portuguese

51 Amitav Ghosh: *The Sea of Poppies*. London: Murray 2008; id.: *River of Smoke*: London: Murray 2011.

52 Beryl Belsky: A Clash of Civilizations: The Ibis Trilogy by Amitav Ghosh. http://beryl-pieces-asia.blogspot.de (accessed October 14, 2013).

53 Gayatri Chakravorty Spivak: The Burden of English. In: Id.: *An Aesthetic Education in the Era of Globalization*. Cambridge, MA / London: Harvard University Press 2012, pp. 35–56.

54 Christopher L. Miller: *The French Atlantic Triangle: Literature and Culture of the Slave Trade*. Durham, NC: Duke University Press 2008.

55 Miller: *The French Triangle*, p. 364.

56 Ineke Phaf-Rheinberger: *The "Air of Liberty". Histories of the South Atlantic Past*. Amsterdam / New York: Rodopi 2008; ead. (ed.): *Historias enredadas. Representaciones asimétricas con vista al Atlántico*. Berlin: Ed. Tranvía 2011.

57 Yves Clavaron / Jean-Marc Moura (eds): *Les empires de L'Atlantique (XIXe–XXIe siècles). Figures de l'autorité impériale dans les lettres d'expression européenne de l'espace atlantique*. Bécherel: Les Perséides 2012.

sea empires, emphasising the outstanding role of the bestseller novel *The Belly of the Atlantic* by the Senegalese writer Fatou Diome.[58] And in another collection of essays, *Shifting the Compass: Pluricontinental Connections in Dutch Colonial and Postcolonial Literature*,[59] the collaborations examine the Dutch- and English-speaking realms, including literature in Afrikaans from South Africa and migrant literatures from Europe.

Obviously, as in historical research, also in literary criticism the intervention of the oceanic world receives more attention. The conceptualisation of this space gives way to a broad disciplinary horizon. A fascinating book that combines history and cultural studies is *Dhow Cultures of the Indian Ocean. Cosmopolitanism, Commerce and Islam*,[60] by the former director of the Zanzibar Indian Ocean Research Institute, the historian Abdul Sheriff. He refers to Braudel's work to organise his own research on the seamen sailing in a specific wooden boat called the dhow:

> The discussion about the development of the dhow has allowed the vehicle to assume a personality of its own. However, we must never forget the people behind the dhow, the lumbermen, who felled the teak forests of Malabar, the dhow builders all around the rim who built dhows, the *nakhodas*, who commanded them, the sailors who manned them, and the passengers who sailed in them across the ocean. Braudel pointed out, "The Mediterranean has no unity but that created by the movements of men, the relationship they imply, and the routes they follow." The same is true of the Indian Ocean. These dhows were commanded and manned by people who were like fish, spending nearly ten months of the year at sea.[61]

This quotation gives insights into the multiplicity of disciplines that are involved when beginning to study and understand the oceanic world, and such deliberations were the point of departure for the organisation of the workshop "Beyond the Line" on 22–23 June 2012 at the Institute of Asian- and African Sciences at Humboldt University, Berlin. The idea was to stimulate, in an interdisciplinary setting, some

58 Fatou Diome: *The Belly of the Atlantic*. London: Serpents Tail 2006. The first French edition *Le ventre de l'Atlantique* (Paris: Éditions Anne Carrière) came out in 2003.

59 Jeroen Dewulf / Olf Praamstra / Michiel van Kempen (eds): *Shifting the Compass: Pluricontinental Connections in Dutch Colonial and Postcolonial Literature*. Newcastle: Cambridge Scholars 2013.

60 Abdul Sherif: *Dhow Cultures of the Indian Ocean. Cosmopolitanism, Commerce and Islam*. London: Hurst 2010.

61 Sheriff: *Dhow Cultures*, p. 99.

contemporary discussions of the above mentioned problems in the context of the southern oceans, in order to erode "the line" debate further. The contributions in this volume express some of the general preoccupations. In the first section, "Studying the Ocean", the subject is life *in* and *on* the water. Georg Berkemer makes a larger theoretical point about the production of historical knowledge with the example of a certain fish (latimeria chalumnae), found off the eastern coast of South Africa in 1938 and previously thought to have been extinct. He discusses the question of "discovery" in history and asks for evidence that might testify to earlier knowledge about the existence of this fish, drawing upon a variety of sources and not only upon written documentation.

Michael Mann shifts the focus toward the forms of everyday life and the regularities of life cycles of people during their time "on the water", be they slaves, coolies, prisoners, or the crew of a ship, and reconstructs their daily behavior, fears, social environment, and occupations. Then Sebastian Prange and Derek Elliot describe the activities of local pirates in the Indian Ocean, who have largely been written out of history until recently. They examine the fragile line between their qualification as criminals or legal entrepreneurs, depending on the historian's evaluation. Margret Frenz focuses on Indian migration in this same ocean after independence from Great Britain in 1947. In Kenya, some Indian migrants became involved in the process of decolonisation, introducing the *swaraj* concept, the demand for self-rule adopted from Mahatma Ghandi, and maintain a more or less continuous contact with developments in India.

In the second section, "Narrating the Ocean", literature is a central concern. Frank Schulze-Engler presents a new interdisciplinary project on "Africa's Asian Options," in which several area studies centers of the University of Frankfurt am Main are involved. The goal is to center on the seminal role of the larger Indian Ocean region in the twenty-first century globalisation processes. Schulze-Engler argues that in particular the African option has to be reviewed, for which the analysis of literary texts offers an important contribution. In this sense, Ute Fendler opens the window to the sea in fictional works from Mauritius, Madagascar, and Mozambique stating that such a comparative approach is quite new. It raises the question to what degree the literary constructions of the sea might contribute to the construction of a shared imaginary of the Indian Ocean—and beyond, with the Caribbean.

Fendler's question forms the bridge to the last two presentations, which address the Atlantic Ocean. Ana Sobral emphasises the role of poetry in Angola since the nineteenth century, drawing a line to Agostinho Neto's anti-colonial poetry in the 1950s and 1960s. In this way, she characterises the maritime space as the "ocean of pain" in remembrance of the slave trade of the past, a tradition that is continued in contemporary poetry and the lyrics of rap-musicians, in which contemporary corruption and misgovernment are crucial themes. In the essay about "oceanic modernity", Ineke Phaf-Rheinberger also analyses the memory of the slave trade in recent literature from Angola and Brazil, departing from the depiction of the port cities. Following Carolyn Cartier's argument in "Cosmopolitics and the Maritime World City,"[62] the entanglements of their cosmopolitan lifestyle develop specific symbolic orientations and multiple interpretations for cultural practices and have significant repercussions in the selected narrative texts.

We would like to express our gratitude to the faculty and staff of the Institut für Asien- und Afrikawissenschaften at Humboldt University for their financial and practical support. We are equally very grateful to the Deutsche Forschungsgemeinschaft for its financial support for the organisation of the workshop and of this book. And last but not least, we thank the contributors for making the publication of this volume possible.

62 Carolyn Cartier: Cosmopolitics and the Maritime World City. In: *The Geographical Review* 89,2 (1999), pp. 278–289.

Bibliography

African Development Bank / African Union (eds): *Oil and Gas in Africa.* Oxford / New York: Oxford University Press 2009.

Ahuja, Ravi: *Pathways of Empire. Circulation, "Public Works" and Social Space in Colonial Orissa, c. 1780–1914.* Hyderabad: Orient Blackswan 2009.

Alpers, Svetlana: *The Art of Describing.. Dutch Art in the Seventeenth Century.* London: Penguin 1989.

Arnold, James A.: From Piracy to Policy: Exquemelin's Buccaneers and Imperial Competition in America. In: *Review* 40,1 (May 2007), pp. 9–20.

Belsky, Beryl: A Clash of Civilizations: The Ibis Trilogy by Amitav Ghosh. http://beryl-pieces-asia.blogspot.de (accessed October 14, 2013).

Benjamin, Thomas: *The Atlantic World. Europeans, Africans, Indians and Their Shared History*, 1400–1900. Cambridge: Cambridge University Press 2009.

Boogaart, E. van den / H R. Hoetink / P. J. Whitehead (eds): *Johan Maurits van Nassau-Siegen 1604–1679. A Humanist Prince in Europe and Brazil.* The Hague: The Johan Maurits van Nassau Stichting 1979.

Bose, Sugata: *A Hundred Horizons. The Indian Ocean in the Age of Global Empire.* Cambridge, MA / London: Harvard University Press 2006.

Boxer, C. R.: *The Dutch Seaborne Empire 1600–1800.* London: Penguin 1990 [1965].

Braudel, Fernand, *Das Mittelmeer und die mediterrane Welt in der Epoche Philipps II,* 3 vols. Frankfurt am Main: Suhrkamp 1990 [*La Méditerranée et le monde méditerranéen à l'époque Philippe II.* Paris: 1949].

Cartier, Carolyn: Cosmopolitics and the Maritime World City. In: *The Geographical Review* 89,2 (1999), pp. 278–289.

Chaudhuri, Kirti N.: *Asia Before Europe. Economy and Civilisation of the Indian Ocean from the Rise of Islam to 1750.* Cambridge: Cambridge University Press 1990.

Clarke, Duncan: *Crude Continent: The Struggle for Africa's Oil Prize.* London: Profile Books 2009.

Clavaron, Yves / Moura, Jean-Marc (eds): *Les empires de l'Atlantique (XIXe–XXIe siècles). Figures de l'autorité impériale dans les lettres d'expression européenne de l'espace atlantique.* Bécherel: Les Perséides 2012.

Curto, Diego Ramada / Francisco Bethencourt (eds): *Portuguese Oceanic Expansion, 1400–1800.* Cambridge: Cambridge University Press 2007.

Davenport, Frances Gardiner: *European Treaties Bearing on the History of the United States and its Dependencies*, 4 vols. Washington, DC: Carnegie Institute 1917–1937, vol. 1: (1716–1815).

Dellai-Schöbi, Katharina: Geheimnisvoller Riesenkalmar. Genanalysen aus der Tiefsee. http://www.nzz.ch/wissen/wissenschaft/genanalysen-aus-der-tiefsee-1.18065139 (accessed February 24, 2014).

DeLoughrey, Elizabeth: *Routes and Roots: Navigating Caribbean and Pacific Island Life.* Honolulu: University of Hawaii Press 2007.

Dewulf, Jeroen / Olf Praamstra / Michiel van Kempen (eds): *Shifting the Compass: Pluricontinental Connections in Dutch Colonial and Postcolonial Literature* . Newcastle: Cambridge Scholars 2013.

Dharampal-Frick, Gita: *Indien im Spiegel deutscher Quellen der frühen Neuzeit (1500–1750). Studien zu einer interkulturellen Konstellation.* Tübingen: Niemeyer 1994.

Diome, Fatou: *The Belly of the Atlantic.* London: Serpents Tail 2006 [*Le ventre de l'Atlantique.* Paris: 2003].

Egerton, Douglas R. / Alison Games / Jane G. Landers / Kris Lane / Donal R. Wright (eds): *The Atlantic World: A History.* Wheeling, IL. Harlan Davidson 2007.

Elliott, John H.: *Empires of the Atlantic World; Britain and Spain in America 1492–1830.* Yale, NH: Yale University Press 2009.

Fisch, Jörg: *Die europäische Expansion und das Völkerrecht. Die Auseinandersetzungen um den Status der überseeischen Gebiete vom 15. Jahrhundert bis zur Gegenwart.* Stuttgart: Steiner 1984.

Freitag, Ulrike / Achim von Oppen (eds): *Translocality. The Study of Globalising Processes from a Southern Perspective.* Leiden: Brill 2010.

Gabaccia, Donna R. / Dirk Hoerder (eds): *Connecting Seas and Connected Ocean Rims. Indian, Atlantic, and Pacific Oceans and China Seas Migrations from the 1830s to the 1930s.* Leiden: Brill 2011.

Ghazvinian,John H.: *Untapped. The Scramble for Africa's Oil.* Orlando: Harcourt 2007.

Gilroy, Paul: *The Black Atlantic. Modernity and Double-Consciousness.* Cambridge, MA: Harvard University Press 1993.

Ghosh, Amitav: *The Sea of Poppies.* London: Murray 2008.

—: *River of Smoke*: London: Murray 2011.

Grotius, Hugo: *The Freedom of the Sea – Or, The Right Which Belongs to the Dutch to Take Part in the East Indian Trade.* New York: Oxford University Press 1916.

Ingram, Bruce S. (ed.): *Three Sea Journals of Stuart Times.* London: Constable 1936.

Klein, Bernard / Gesa Mackenthun (eds): *Das Meer als kulturelle Kontaktzone. Räume, Reisende, Repräsentationen.* Constance: ULV 2003.

—: (eds): *Sea Changes. Historicizing the Ocean.* New York / London: Routledge 2004.

Klein, Herbert S.: *The Atlantic Slave Trade.* Cambridge: Cambridge University Press 2002.

Lago, Pedro Corrêa do / Bia Corrêa do Lago: *Frans Post 1612–1680. Obra Completa.* Rio de Janeiro: Capivara Editora 2006.

Lefebvre, Henri: *The Production of Space*, trans. by Donald Nicholson-Smith. Oxford / Cambridge, MA: Basil Blackwell 1991 [*La production de l'espace.* Paris: 1974].

Lewis, Martin W.: Dividing the Ocean Sea. In: *The Geographical Review* 89,2 (1999), pp 188–214.

Lewis, Martin W. / Karen Wigen: A Maritime Response to the Crisis in Area Studies. In: *The Geographical Review* 89,2 (1999), pp. 161–168.

Mann, Michael: *Sahibs, Sklaven und Soldaten. Geschichte des Menschenhandels rund um den Indischen Ozean.* Darmstadt: WBG 2012.

Mattingly, Garrett: No Peace Beyond What Line? In: *Transactions of the Royal Historical Society* 13 (1963), pp. 145–162.

McPherson, Kenneth: *The Indian Ocean. A History of People and the Sea.* Delhi et al.: Oxford University Press 1998.

Meredith, Martin: *The State of Africa. A History of Fifty Years of Independence.* The Free Press 2005.

Miller, Christopher L.: *The French Atlantic Triangle: Literature and Culture of the Slave Trade.* Durham, NC.: Duke University Press 2008.

Muguet, Marc: *Après Cousteau. Le futur de l'Océan.* Paris: Éditions Yago 2010.

Mukherjee, Rila (ed.): *Oceans Connect: Reflections on Water Worlds Across Time and Space.* Delhi: Primus Books 2012.

Müller, Leos / Philipp Robinson Rössner / Toshiaki Tamaki (eds): *The Rise of the Atlantic Economy and the North Sea-Baltic Trade, 1500–1800.* Stuttgart: Steiner 2011.

Osterhammel, Jürgen / Petersson, Niels P.: *Geschichte der Globalisierung. Dimensionen, Prozesse, Epochen.* Munich: C. H. Beck 2003.

Paasman, Bert A. N.: *Nederlandse Literatuur en slavernij ten tijde van de Verlichting.* Leiden: Nijhoff 1984.

Parker Brienen, Rebecca: *Visions of Savage Paradise. Albert Eckhout, Court Painter in Colonial Dutch Brazil.* Amsterdam: Amsterdam University Press 2006.

Pearson, Micheal N.: *The Indian Ocean.* London / New York: Routledge 2003.

Phaf-Rheinberger, Ineke: *The "Air of Liberty". Histories of the South Atlantic Past.* Amsterdam / New York: Rodopi 2008.

Phaf-Rheinberger, Ineke (ed.): *Historias enredadas. Representaciones asimétricas con vista al Atlántico.* Berlin: Ed. Tranvía 2011.

—: *Vision* and the Representation of Africans – On Historical Encounters between Science and Art. In: *History of Philosophy and the Life Sciences* 35 (2013), pp. 53–61.

Pomeranz, Kenneth / Steven Topik: *The World that Trade Created. Society, Culture, and the World Economy.* New York / London: Shape 2006.

Pratt, Mary Louise: *Imperial Eyes. Travel Writing and Transculturation.* London / New York: Routledge 1992.

Ptak, Roderich: *Die maritime Seidenstraße. Küstenräume, Seefahrt und Handel in vorkolonialer Zeit.* Munich: Verlag C. H. Beck 2007.

Rothermund, Dietmar / Susanne Weigelin-Schwiedrzik (eds): *Der Indische Ozean. Das afro-asiatische Mittelmeer als Kultur- und Wirtschaftsraum.* Wien: Mandelbaum 2004.

Schmitt, Carl: *Land und Meer. Eine weltgeschichtliche Betrachtung.* Stuttgart: Klett-Cotta 2008 [1942].

Schmitt, Eberhard (ed.): *Indienfahrer 2. Seeleute und Leben an Bord im Ersten Kolonialzeitalter.* Wiesbaden: Harrassowitz 2008.

The Sea in Nineteenth-Century English and American Literature. http://www.enotes.com/topics/sea-nineteenth-century-english-and-american/critical-essays (accessed October 14, 2013)

Sherif, Abdul: *Dhow Cultures of the Indian Ocean. Cosmopolitanism, Commerce and Islam.* London: Hurst 2010.

Sobel, David: *Longitude. The True Story of a Lone Genius who Solved the Greatest Scientific Problem of his Time.* Harmondsworth et al.: Penguin 1995.

Spivak, Gayatri Chakravorty: The Burden of English. In: Id.: *An Aesthetic Education in the Era of Globalization.* Cambridge, MA / London: Harvard University Press 2012.

Tvedt, Terje et al. (eds): *A History of Water*, 3 vols. London: Tauris 2006–2011.

Villalón, Leonardo / Peter von Doepp (eds): *The Fate of Africa's Democratic Experiments. Elites and Institutions.* Bloomington: Indiana University Press 2005.

Wasafiri 66 (2011): Indian Oceans.

Weltmeere steuern auf Kollaps zu. http://www.stern.de/wissen/natur/forscher-schlagen-alarm-weltmeere-steuern-auf-kollaps-zu-1697922.html (accessed February 24, 2014).

Williamson, James A.: *Hawkins of Plymouth. A New History of Sir John Hawkins and of the Other Members of his Family Prominent in Tudor England.* 2nd ed. London: Black 1969.

I.
Studying the Ocean

The Seaman and the Fish

Discoveries beyond the Line

Georg Berkemer*

Introduction

Historians deal in facts. Facts, as people who know their Latin can tell us, are 'made' (factus) by people from the past whose texts were subject to the continuities and contingencies of textual transmission and the power politics in the production of historical knowledge. Historians use source materials, namely the texts still available in their time to construct narrations of the past. Historians are scholars who ply their trade safely in the back area, sifting through the debris after the armies have left the battlefield and the explorers have finished their travelogues. The historian's field of discovery is the archive, a place where the general public expects to encounter mainly dust and brittle tomes of bygone ages.

What can a historian say about the discovery of a fish, highly remarkable as the species in question may be? First, historians create tales of past events in a way that later generations may be able to draw their own conclusions from those tales. The way these historical narratives are constructed differ from writer to writer, age to age and motive to motive, as Hayden White has demonstrated.[1] Such tales are, as a

* I wish to thank Hanna Werner, Johannes Waligora and the editors for their help and constructive criticism.

1 Hayden White: The Narrativization of Real Events. In: *Critical Inquiry* 7 (1981), pp. 793–798; id.: The Question of Narrative in Contemporary Historical Theory. In: *History and Theory* 23 (1984), pp. 1–33; id.: *Tropics of Discourse. Essays in Cultural Criticism.* Baltimore: Johns Hopkins University Press 1992.

basic tenet of post-modern critique holds, never 'objective', and various authors stress the fact that academic discourses are always fraught with politics and power discourse.[2] However, a certain concern about a genuine methodology in the treatment of one's data, an attitude which one may call 'honesty', shared by all participants in the discourse has to be taken for granted.

There is a drawing of a fish in a navigator's journal, dated 1598, which intrigued me. And it inspired me to tell my own story of a failed attempt to overcome the dividing line between the past of sixteen hundred AD, the modern almost-present of 1938, and the post-colonial present. I asked myself: did I by accident make a find in this book from the age of discovery that amounts to a small sensation in the world of fish researchers as well as historians? Or to be more concrete: the coelacanth (Latimeria chalumnae) was (re-)discovered in 1938. Could it be possible that this unexpected discovery had happened much earlier already, but was first ignored and then entirely forgotten?

Before I tell the story of the coelacanth, however, some theoretical considerations are in order. There is a line between two types of knowledge as seen from the point of view of the Western public. They are very different in status. One comprises such common knowledge as needed by laymen for their daily work. It is the knowledge used for instance by fishermen and sailors. The other is scientific knowledge. Western modernity, if one is for a minute to assume the somewhat uncritical 'modern' outlook at such a modernity itself, has achieved the formalization and mathematization of knowledge, most successfully of natural phenomena. This was among other things the consequence of the comprehensive collection and musealization of human and natural history, and more recently the digitalization of this information. Behind this effort is "the image of a primordial truth fully adequate to its nature, and it necessitates the removal of every mask to ultimately disclose an original identity."[3] In other words, 'things' are taken out

2 Michel Foucault: Truth and Power. In: Id.: *The Foucault Reader. An Introduction to Foucault's Thought*, ed. by Paul Rabinow. London: Penguin Books 1991, pp. 51–75; Hayden White: The Politics of Historical Interpretation. Discipline and De-Sublimation. In: *Critical Inquiry* 9 (1982), pp. 113–137.

3 Michel Foucault: Nietzsche, Genealogy, History. In: Id.: *The Foucault Reader. An Introduction to Foucault's Thought*, ed. by Paul Rabinow. London: Penguin Books 1991, pp. 76–100, here p. 78.

of their context and re-contextualized in the discourse of an expert culture.

Let us call these achievements of Western production of knowledge—in the spirit of Derrida and Foucault—the contents and procedures of 'The Archive'. This deliberate essentialization I use to denote all the repositories of knowledge which Western civilization created, and still creates, in order to store information about itself and the world before and outside. In this view of the world, anything not yet included in the archive is seen as 'undiscovered'. The archive also contains the rules used to determine what information to include or exclude, the way potential knowledge should be categorized and utilized, and the rules about who is allowed access to which information.[4] As in the case of historical narration, the question of 'truth' arises here:

> The perfect Archive is ex hypothesi an evidence which cannot lie to us: we may through laziness or other imperfection of our own misinterpret its statements or implications, but itself it makes no attempt to convince us of fact or error, to persuade or dissuade: it just tells us. That is, it does so always provided that it has come to us in exactly the state in which its original creators left it. Here then, is the supreme and most difficult task of the Archivist—to hand on the documents as nearly as possible in the state in which he received them, without adding or taking away, physically or morally, anything: to preserve unviolated, without the possibility of a suspicion of violation, every element in them, every quality they possessed when they came to him, while at the same time permitting and facilitating handling and use.[5]

Already in the 19th century, rules for classification of species had been established by expert committees[6] and further developed into

4 There seems to be no such word as 'valeur archivistique' oder 'Archivwürdigkeit' in English (cf. Simone Chiquet: Was heisst eigentlich archivwürdig? In: *Schweizerische Zeitschrift für Geschichte = Revue suisse d'histoire = Rivista storica svizzera* 51,4 (2001): Überlieferungsbildung und Bewertung = Evaluation et formation des sources archivistiques, pp. 470–486). An approximate translation would be 'eligibility to be archived'. This power to decide what is to be labeled as such shows on a very basic level of daily archival work how collective memory is determined by power discourses.

5 Hilary Jenkinson: Reflections of an Archivist. In: *Contemporary Review* 165 (1944), pp. 355–361, quoted in: Anne Gilliland-Swetland: Enduring Paradigm, New Opportunities. The Value of the Archival Perspective in the Digital Environment. Washington, D.C.: Council on Library and Information Resources 2000. http://www.clir.org/pubs/reports/pub89/contents.html (accessed August 20, 2013).

6 Earliest was Alphonse de Candolle: Laws of Botanical Nomenclature Adopted by the International Botanical Congress Held at Paris in August 1867; Together with an Historical Introduction and Commentary by Alphonse de Candolle. In: *The American Journal of Science and Arts, Series II* 46 (1868), pp. 63–77 (First published in French,

the present-day databases.[7] Similar systems of types exist in mineralogy. The type material is usually stored for reference in a museum of national importance. All such systems of type formalization and nomenclature enable experts to deal with the basic information of their field according to the paradigm they adhere to. All archival materials together with their meta-data and the principles of archival practices were, as maintained by archival experts, developed to "transcend mere information, and mere information management" with the final aim "to lead others to seek, 'knowledge' and meaning among the records in their care".[8] The archive, as Derrida and Foucault warn us is also a thing of power.[9] What items are allowed to be incorporated, at what order, and to what purpose, is a matter of contention. Those who have access to its contents, are entitled to speak about them, are in charge of the archive's order or control access to it (of both its new entries and its users), gain status and influence. Knowledge is power, and the archive, like Foucault's somewhat ironical definition of 'truth' is "as system of ordered procedures for the production, regulation, distribution, circulation and operation of statements".[10]

The 'Discovery' of the Coelacanth

In comparison with the historiography of the archive and its relation to 'truth', the story of the discovery of the coelacanth as told by fish experts is a straightforward one:[11] It was in 1938, on board a

Genève / Bâle / Paris: H. Georg et J.-B. Bailliere 1867; Reprinted from the English translation published by L. Reeve and Co., London 1868).

7 For zoology the International Commission on Zoological Nomenclature was active since 1895. Cf. iczn.org/content/about-iczn (accessed January 17, 2014). Cf. also Neal L. Evenhuis: A Compendium of Zoological Type Nomenclature: a Reference Source. Electronic document, 2008. http://hbs.bishopmuseum.org/publications/pdf/bm-tp41.pdf (accessed September 18, 2013). An important site specialized on fishes is fishbase.org (accessed January 17, 2014).

8 Terry Cook: From Information to Knowledge: An Intellectual Paradigm for Archives. In: *Archivaria* 19 (1984), pp. 28–49, quoted in: Gilliland-Swetland: Enduring Paradigm. http://www.clir.org/pubs/reports/pub89/contents.html (accessed August 20, 2013).

9 Jacques Derrida: *Archive Fever. A Freudian Impression*. Chicago: University of Chicago Press 2008; Michel Foucault: *The Order of Things*. London / New York: Routledge 2002.

10 Foucault: Truth and Power, p. 74.

11 There is a well maintained website dedicated solely to Coelacanth research—http://www.dinofish.com (accessed December 24, 2013)—which is in close cooperation with

Fig. 1
Dr. Latimer
with the first specimen
of Latimeria chalumnae.

fishing vessel from the South African port of East London that Henrik Goosen, the trawler's captain, caught sight of a strange creature, large and blue, among the catch. Marjorie Courtenay Latimer, curator at the local museum, secured the curious specimen for detailed description. It was a fish, but Dr. Latimer had never seen such a creature before. The only comparable fish she could find was an unlikely one, namely to species of the Coelacanthiformes order. This order contained several genera of coelacanths, all of which had become extinct by the end of the Jurassic period (200–145 mio. years ago). Coelacanths were known to have been lobe-finned fishes and were more closely related to recent lungfish and reptiles than to the common modern ray-finned fishes.

Failing to find any more plausible descriptions of her fish in any of her books, Dr. Latimer sent the unknown find to a taxidermist as the only means of preservation, and contacted a friend, the South African Professor James L. Brierley Smith. It was Smith who confirmed the creature as a living relative to the fossil order of the coelacanths, and named the fish Latimeria chalumnae in honour of Marjorie

the South African Institute of Aquatic Biodiversity. Cf. http://www.saiab.ac.za/home.htm (accessed December 24, 2013). See also Penny Haworth (ed.): *Building on the South African Coelacanth Legacy. African Coelacanth Ecosystem Programme 2001–2013. 75th Anniversary 1938–2013*. Grahamstown: The South African Institute for Aquatic Biodiversity / The African Coelacanth Ecosystem Programme 2013. http://www.saiab.ac.za/uploads/files/saiab_acep_report_2013_website.pdf (accessed November 19, 2013), one of the many to-be-expected publications for the 75th anniversary of the 'discovery' first coelacanth.

Courtenay-Latimer[12] and after the waters in which it was found.[13] This first specimen is still on display in the East London museum.

The discovery made the news all over the world, or was "head-lined into human consciousness with its 'discovery' alive in 1938"[14]. A second coelacanth was captured in 1952 off the coast of the island of Anjouan in the Union of the Comoros, and flown to South Africa by J. L. B. Smith. The specimen's dorsal fin was missing and its tail fin deformed.[15] Since then, several more Latimeriae were caught by fishermen or discovered by scientists in local markets. More recently, the animals have been filmed in their natural habitats, and a second species was found in Indonesia.[16] It is now clear that the two Latimeria species have a wide range along the coast of Eastern Africa and in the Indian Ocean. However, the coelacanths are at risk of extinction due to human influences. The African species Latimeria chalumnae is listed as "critically endangered", while the second species (Latimeria menadoensis or Sulawesi coelacanth) is listed as "vulnerable" by the International Union for Conservation of Nature (IUCN).

By 2010, the story of the coelacanth is told on various sites in the Internet. Content and even wording is often so similar that authorship of the texts can no more be established with any certainty.[17] All

12 J. L. B. Smith: A Living Fish of Mesozoic Type. In: *Nature* 143 (1939), pp. 455–456 .

13 The fossil as well as the living species belong to the order Coelacanthiformes. There are several genera of extinct species within the family of the Coelacanthidae, among them the genus Coelacanthus. The family of the recent Latimeriidae is one among several others within the order Coelacanthiformes. It consists of the single genus Latimeria.

14 http://www.dinofish.com (accessed January 17, 2014).

15 J. L. B. Smith: The Second Coelacanth. *Nature* 171 (1953), pp. 99–101.

16 Mark V. Erdmann / L. Caldwell / M. Kasim Moosa: Indonesian 'King of the Sea' Discovered. In: *Nature* 395 (1998), p. 335; L. Pouyaud / S. Wirjoatmodjo / I. Rachmatika / A. Tjakrawidjaja / R. Hadiaty / W. Hadie: Une nouvelle espèce de coelacanthe: preuves génétiques et morphologiques: A New Species of Coelacanth. In: *Comptes Rendus de l'Académie des Sciences, Series III – Sciences de la Vie* 322 (1999), pp. 261–267. For a complete list of all specimens up to 2012 cf. http://www.dinofish.com/recenthistory.html. Another recently published list is Rik Nulens / Lucy Scott / Marc Herbin: *An Updated Inventory of All Known Specimens of the Coelacanth, Latimeria spp.* Grahamstown: South African Institute for Aquatic Biodiversity / African Coelacanth Ecosystem Programme 2011 (= Smithiana Special Publication, vol. 3). http://www.saiab.ac.za/uploads/files/spec._pub._3_-_coelacanth_inventory.pdf (accessed December 13, 2013).

17 Cf. the dozens of sites on Wikipedia in various languages, drawing heavily

Fig. 2: Specimen of Latimeria chalumnae in the Natural History Museum, Vienna, Austria, caught 1974 next to Salimani / Comoro Islands.

this indicates the activities of a 'community' of enthusiasts within the Western archive who have access to specialized literature and Internet sites.

The Internet sites created by them mention local knowledge about the Latimeriae which never established itself in the form of a written archive—or nowadays an electronic database—which was the only proper way to present information and accumulate knowledge according to exclusivist western public and scientific norms. The inclusion of this type of knowledge and the voices of those who participate in this type of knowledge production demonstrates the power discourse about the archive and recent changes in the rules of access. What people from outside the established circle of experts knew about the Latimeria species was irrelevant as long as their knowledge could not be incorporated by themselves into the mainstream or dominant discourse of western media, scientific journals and historical records.

Most likely, fishermen and visitors to local fish markets knew the animals before they became available for description by the scientific community. Presently, Latimeriae are an unusual catch in their home waters, but they are by no means a sensational find. As such, they feature in

from each other as well as from http://dinofish.com and local sites such as http://www.saiab.ac.za/newsroom/building-on-the-south-african-coelacanth-legacy.htm (accessed June 3, 2013) and http://www.sahistory.org.za/dated-event/fishing-trawler-south-african-coast-catches-coelacanth (accessed June 3, 2013). There are cross references to more scientifically inclined sites such as http://fishbase.org (accessed June 3, 2013) and publications in Internet journals, especially back issues of *Nature*, the journal in which the discovery was originally published.

Fig. 3: Garbage found in the stomach of a specimen caught in Indonesia in 2011.

the local tales and histories of the fishermen and their customers at the markets in Africa, Indonesia and the Comoros. For that reason, *dinofish.com* uses the words 'discovery' and 'discovered' in quotation marks.[18] What has been 'discovered' by the scientific community since 1938 was not only the fish as biological entities, but also cultural techniques concerning their catch and their preparation, their valuation as a food source, names in different languages, etc. From all this material follows that the local fishing people knew about the coelacanths long before Latimer and Smith 'discovered' them.

18 http://dinofish.com (accessed January 17, 2014) is a site run by coelacanth enthusiasts who lay emphasis in the conservation of the latimeria species and the dissemination of knowledge about them. They also stress the fact that in order to preserve the environment of these fish, the local population has to be included in their preservation efforts. Therefore, they stress that the coelacanths were known, and therefore 'discovered' by them, rather than the scientific community. Other popular literature still does not hesitate to use this misleading term (Samatha Weinberg / Andrea Stumpf / Gabriele Werbeck: *Der Quastenflosser. Die abenteuerliche Geschichte der Entdeckung eines seit siebzig Millionen Jahren vermeintlich ausgestorbenen Tieres*. Berlin: Argon 1999).

I do not intend to discuss here the local forms of knowledge. I want, however, ask whether there were or could have been older, alternative ways for these fish to enter the historical archives. I want to ask the following question: Were the Latimeria species—maybe accidentally—included into the western archive of discoveries before *Nature, dinofish.com* and similar media created specialized archives for them? The assumption that Latimeria was known to western people before 1938 was exactly the basis for the coelacanth hoax of the 1960s.

The Coelacanth Hoax

Soon after the Latimeriae were included into the western archive in 1938, attempts to make a profit from it began. There are coelacanth coins from South Africa, coelacanth stamps (Kuwait, Mozambique, South Africa), and various items designed for the souvenir market.[19] Around 1965, two silver fish pendants looking very much like miniature coelacanths were offered for purchase on the antiques market in Spain. The pendants were made in a traditional style and described as religious votive figures from Spanish churches. The time of manufacture was estimated between the seventeenth and the late nineteenth century. If this was true, it would change our perception about the 'discovery' of the coelacanth. It would have implied that the Latimeriae were known to Hispanic seafarers long before the 'discovery' for science in 1938. It was speculated that the specimens represented here originated either from the Mediterranean, subtropical Atlantic, or subtropical Pacific, and that the silver fish pendants might even indicate the existence of other surviving species of Coelacanthiformes.[20]
New studies reveal that the silver fish artefacts are indeed coelacanths, but were in fact produced more recently. It was demonstrated that the unknown makers faithfully copied the damaged fins and head sections

19 http://www.coelacanthwatch.com (accessed June 5, 2013) is such a site. While professing to "promote this eco-marine heritage to the local and international industry by marketing Coelacanth products which have been designed and manufactured by local crafts people", their interest in commercial success seems to be predominant.

20 D.P. De Sylva: Mystery of the Silver Coelacanth. In: *Sea Frontiers* 12 (1966), pp. 172–175; Michael N. Bruton / Sheila E. Coutouvidis / Jean Pote: Bibliography of the Living Coelacanth *Latimeria chalumnae*, with Comments on Publication Trends. In: John A. Musick / Michael N. Bruton / Eugene K. Balon (eds): *The Biology of Latimeria chalumnae and Evolution of Coelacanths*. Dordrecht / Boston: Kluwer Academic Publishers 1991, pp. 403–433.

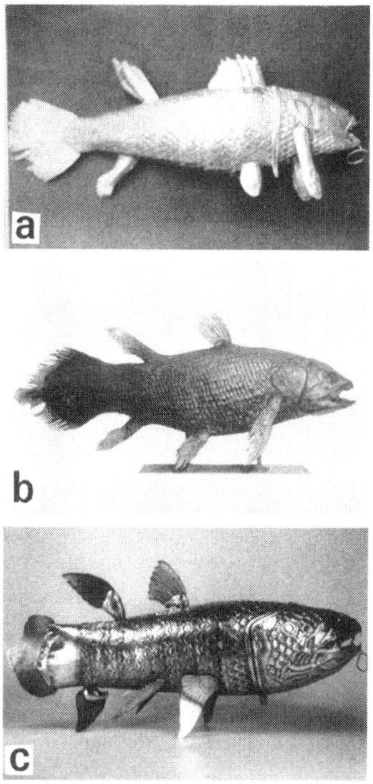

Fig. 4
"a The 1964 coelacanth artefact of Bilbao, Spain;
b the holotype of 1938 displayed in the Natural History Museum in East London, South Africa;
c the 1965 coelacanth pendant of Toledo, Spain."

of the first Comoran coelacanth of 1954. Fricke and Plante prove convincingly that these artefacts cannot have been made before that date and are simply modern adaptations of a traditional style and an attempt to cash in on the coelacanth hype.[21]

So, it remains an established fact that the archive does not contain any information on the Latimeriae from before 1938. As an illustration of this, here a quote by another coelacanth expert, James Atz:

> The most recent half millennium of its existence must have been spent not too far from people who plied their vessels one to two hundred meters above its natural habitat and who occasionally actually caught it by long-lining—although *nothing ever seems to have been recorded* about this hit-or-miss association.[22]

21 Hans Fricke / Raphael Plante: Silver Coelacanths from Spain are not Proofs of Pre-scientific Discovery. *Environmental Biology of Fishes* 61 (2001), pp. 461–463. Another hoax is reported on http://www.dinofish.com/contr.htm (accessed November 22, 2013).

22 James W. Atz: The Lure of Latimeria. In: *Environmental Biology of Fishes* 61 (2001),

The time frame given in this quotation is historically incorrect and too much oriented towards European shipping, for ocean-going ships plied the seas of the East African coast and Madagascar from at least 800 A.D., and one is probably not mistaken to add at least one millennium more to this figure in Latimeriae's Asian habitat in Indonesia. Atz is probably right stating that no records of sightings of a Latimeria before 1938 are known to researchers from the Western archive. This, for a historian, calls for a few critical remarks: we can be fairly sure that there were humans who have seen and even touched a Coelacanth before 1938, simply inferring from the fact that the fish takes a hook and has been seen after 1938 at local markets all over the region of its Western Indian Ocean habitat, in Indonesia and maybe even in the Pacific Ocean.[23] Thus, there is no reason to assume that this was not the case before the 'official' discovery by Latimer and Smith in 1938. However, none of these people had any intention to do more with the fish than sell or eat it. The problem, therefore, is not the fish, but the archive and its segmentation. As already stated, traditional fishing people did not have access to the archive, ichthyologists rarely look into historical records, and historians do not concern themselves with fish. This separation is unfortunate, because by bringing together both sections of the archive, a new aspect of the problem comes to light.

The Voyage of Heyndrick Dirrecksen Jolinck

In 1598, the second Dutch expeditionary fleet was sent into the Indian Ocean under the command of Jacob Cornelisz van Neck (1564–1638) and Wybrandt Warwijk (1566 / 1570–1615). The objective was to collect intelligence about shipping routes around Africa to the East and

pp. 347–351, here p. 351.

23 In the context of population sizes, http://www.dinofish.com/pop.htm (accessed March 8, 2013) discusses the frequency of catches of coelacanths: "Certainly, the Comoro islands are the only place where regular annual catches of 6–8 coelacanths occur at Grand Comoro and 4–5 at Anjouan. But from time to time coelacanths turned up elsewhere: the one trawled off East London, in 1938; the one trawled off Maputo, Mozambique on August 11, 1991; the four netted off Madagascar in 1995 / 1997 / 2001 - along with other rumored and newly reported Madagascar catches; and now the Sodwana, South Africa, Kenya / Tanzania / Zanzibar catches." On the same web page further catches are mentioned: "Finally, Jerome Hamlin, dinofish.com author and webmaster has found evidence that coelacanths are being caught in the Solomon Islands of the South Pacific! In 2010, a Fukushima Aquarium expedition confirmed coelacanths as far east as Papua Indonesia."

Dusdanige vis sonder steert gesien omtrent 38 mijlen vant grote eilandt Madagascar op die hoochte van 28 gr.; wit onder sijn buijck.

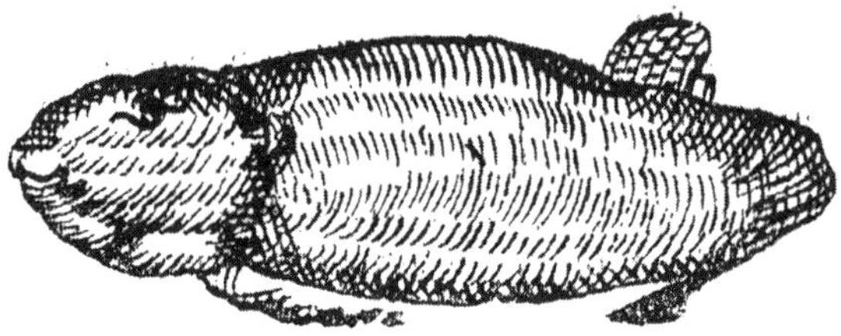

Fig. 5: Reproduction of Jolinck's sketch of the fish as found in his manuscript of 1598 and in the printed version of 1947 (Translation: "This horrible fish without a tail seen more than 38 miles off the large island of Madagascar at the latitude of 28 degrees; white underneath his belly.")

the spice market there, and to secure a load of spices at Ternate, in the spice islands. The voyage was a huge commercial success, bringing a return of 250–400 per cent for its backers, and a large amount of information in the form of logs and journals written by various officers and merchants. The original documents are kept at the Nationaal Archief at Den Haag and have been published in seven parts between 1938 and 1951. One of the navigators, by name of Heyndrick Dirrecksen Jolinck, included several interesting drawings of sea animals in his journal, so for instance an angel shark.[24]

And then, on page 56 of the printed edition, there is this: Is this a coelacanth? At first glance, it does not seem to be the case.[25] The telltale fleshy fins are missing and only one dorsal fin is depicted. From Jolinck's description on the same page in his journal, however, it seems that the "horrible fish" had the right size. It was unable to flee when the crew tried to harpoon it and "wallowed like a ship". The crew

24 Heyndrick Dirrecksen Jolinck: A: Het Journaal van H. D. Jolinck. In: Jacob Cornelisz van Neck / Wijbrand van Warwijck / Johannes Keuning (eds): *De Tweede Schipvaart der Nederlanders naar Oost-Indië onder Jacob Cornelisz. van Neck en Wybrandt Warwijck 1598–1600. Journalen, Documenten en andere Bescheiden uitgegeven en toegericht door J. Keuning. Deel V, Eerste Stuk: De Reis naar Ternate.* 's-Gravenhage, Martinus Nijhoff 1947 (Werken uitgegeven door de Linschoten-vereeniging 50).

25 A look by the author of these lines at the original at the Nationaal Archief in Den Haag in 2004 revealed nothing new. The reproduction is just as good as the original sketch.

could not pull it out of the water and when it slipped away Jolinck could not do more than sketch it from memory "according to my best ability". One should note that Jolinck looked down at the fish from several metres above the ocean surface and tried to depict something which he could hardly perceive in detail, namely the side and belly of the creature which either moved by itself or was tossed about by the waves.[26] A fish expert told me in 2010 that part of the fish could be missing. So maybe it had a tail which was eaten away? This would explain why this specimen was adrift at the surface of the ocean, away from its usual habitat in the dysphotic zone, about 100 to 200 metres below the surface of the ocean.

On the other hand, the location given by Jolinck, "38 miles off the large island of Madagascar" is coelacanth country for sure,[27] and when one compares this sketch with the attempt by Dr. Latimer to catch the likeness of her coelacanth in 1938, certain similarities are striking.[28] Another possible candidate could be the Slender Sunfish (Ranzania laevis) which has no tail fin, but in this case the Sunfish's very characteristic dorsal and anal fins would be missing.

However, all this does not prove anything. There still remains the question: Did Jolinck find the coelacanth? It seems that in our case the historian can only provide the materials available to her / him as a specialist in archival work, and ask the crucial questions. It has to be left to others to decide whether the coelacanth was 'discovered' and forgotten by Europeans centuries before Marjorie Latimer found it again in 1938 in the nets of a South African trawler. What the historian can do, however, is keep an open mind when sifting the information stored in

26 Coelacanths cannot survive at the surface of the sea due to their adaptation to the temperature and pressure at the depth of its habitat. http://dinofish.com (accessed November 13, 2013).

27 "Madagascar was long thought to be a potential site for coelacanth populations (Particularly by researcher Robin Stobbs). Scientifically confirmed catches began in the mid 1990's. The 4th 'official' Madagascan coelacanth was caught by accident in a gill net north of Toliara in mid-March, 2001. This specimen was a female about 1.8 m in length and 80kg. It was caught at 100m depth. In 1995, and again a year later, two coelacanths were netted near St Augustin. A third was later caught in the same area but cut up for bait before being examined by scientists! By now, a half dozen more have been brought up. So far it looks as if the Madagascar coelacanths are the same species as the Comoran ones and may be 'driftees.'" http://dinofish.com (accessed November 13, 2013).

28 See Latimer's sketch on p. 44, Fig. 6.

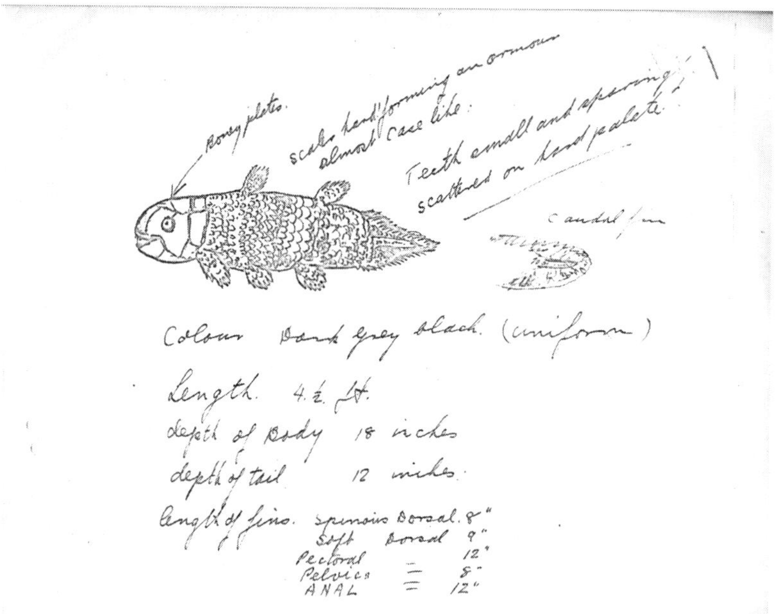

Fig. 6: Dr. Latimer's sketch of the coelacanth, drawn in 1938 for J. L. B. Smith.

old records and archives. Production of knowledge can only be successful when we use the information available there to ask the right questions and formulate feasible hypotheses.

So this essay can only end with the disquieting note that again the Western archive, and especially that of the historian, is neither complete nor objective. Its 'facts' are, like the contents of any other repository of knowledge, good for debate and historical narrative. And 'facts' do not lead us to any kind of truth, since "[t]ruth is undoubtedly the sort of error that cannot be refuted because it was hardened into an unalterable form in the long baking process of history".[29]

This paper shows that there is material in the archive which enables us to take a second look at the question as to how and when the coelacanth was 'discovered' or actually re-contextualized into and within the western archive. The initial find of 1938 is to be understood in the then current Darwin discourse. While within this context the

29 Foucault: Nietzsche, p. 79.

essentialist category of 'the fish' was already slowly giving way to more accurate evolutionist ways of classification, it is a valid question for historians to try and find the context of previous discourses. Today, discourse about the coelacanth occurs mainly in two contexts: the now mainstream theory of biological evolution with its various definitions of species (cladistic, typological, morphological etc.) and genomic research, and second the environmentalist discourse. A third, older context, which may be called subaltern, can also be observed as a subtext in these discourses: the local one, which slowly integrates itself mainly into the environmentalist debates.

In a pre-Darwinian context, ideas were current which could have had an impact on the way how Jolinck himself and later readers of his text might have evaluated his discovery. In the 16th century, almost anything that lives in the water, carp, shark, hippopotamus, molluscs or beaver could be subsumed under the category of 'fish' (and therefore consumed during Lent). The essentialist idea of the species as pre-existing[30] from day one implies that each animal belongs to a species and will be 'discovered' in due time. There was, however, one necessary rule for such classifications: the type specimens, meeting the Aristotelian laws of natural forms according to which species were described, had to be the representative exemplars of a group of individual animals. In other words: a species could only be said to exist if there were more than just one individual, and all carried the common traits that made them as species. Otherwise, the single find of something like a malformed individual, something chimera-like—a monster—had to be put in a category outside this ordered universe. Jolinck's "horrible fish" did not meet the expectations of the day concerning a 'proper' species of fish, and it did occur merely once.

Other strange creatures encountered by the 1598 expedition, such as the dodo and the angel shark were unexpected, but they were species. Funny ones maybe, which lived in waters and on islands beyond the line, but to be encountered predictably in their natural ranges, and therefore (before extinction) material fit for the inclusion in the western archive. Their observations were in accordance with the discourse

30 "It has to be a fish if there is the idea of a fish in the first place!" (BBC Two TV show *QI* (Quite Interesting), D: Ian Lorimer (QI series H episode 3 Hoaxes – Is there such a thing as a fish-.mp4, 1:49); file viewable under http://www.youtube.com/watch?v=NdNEN4rUOTA (accessed November 11, 2013).

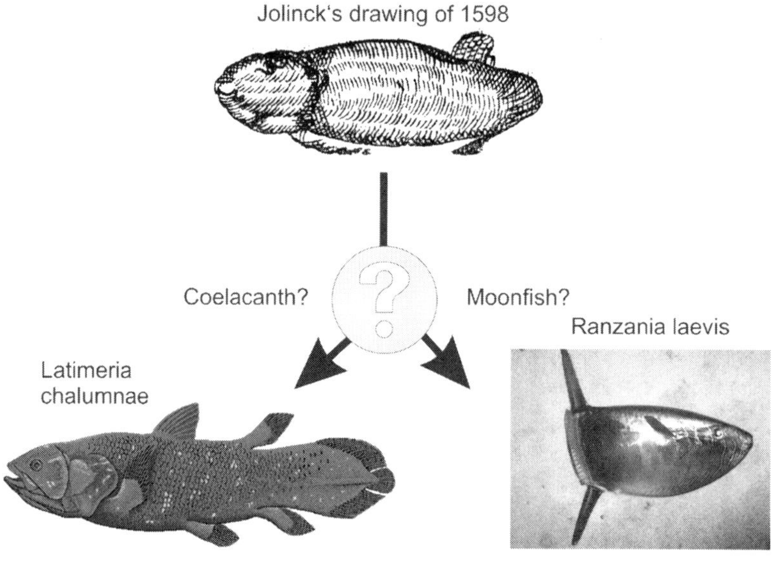

Fig. 7: Possible interpretations of Jolinck's sketch..

of his time—that of natural philosophy. In the case of the coelacanth, the contexts of the potential 'discovery' of 1598 and the actual one of 1938 which show the change of paradigm in which they occurred. In 1938, anything new would be tested against the fossil record and the latest version of the theory of evolution in which change itself was the norm, and anything newly discovered was expected to be explainable within the current paradigm.

In 1598, no such discourse existed. As a ship's pilot charged with the maintenance of a journal, Jolinck neither had the evidence at hand nor the authority to create such a context. The cadaver was lost, the suitable paradigm for further research non-existent. When he wrote and sketched the "horrible fish", this exemplar did not fit into the Aristotelian essentialist idea of animal species which was the current paradigm before the idea of biological evolution gained ground. Here, sightings of monsters were part of the occupational hazards for those who traveled the seas through uncharted waters. Such creatures, however, were not in essence part of any 'species'. They were either seen as aberrations in God's creation, or part of that yarn one had to expect from a jack-tar. They were not type specimens for the classification of an unknown taxon of organisms. It is therefore quite possible that the coelacanth, like the basking or megamouth shark (Megachasma pelagios)[31], the angel sharks (Squatina spp.), the giant oar fish (Regalecus glesne), the giant squids (Architeuthis spp.) and the colossal squid (Mesonychoteuthis hamiltoni) are monsters which did not fit the Aristotelian essentialist morphology of species. The only result of Jolinck's activities was that his journal was taken into the archival record of the expedition and is now preserved at The Hague.

It is the task of the historian to narrate these re-contextualisations over time. He / she seeks connections between the various discourses over social and professional boundaries and over time and cultures.

31 This extraordinary creature was 'discovered' in 1976 off the island of O'ahi, Hawai'i. Cf. Tim M. Berra: Some 20th Century Fish Discoveries. In: *Environmental Biology of Fishes* 50 (1997), pp. 1–12 and Leighton R. Taylor / L. J. V. Compagno / Paul J. Strutsaker: Megamouth – A New Species, Genus, and Family of Lamnoid Shark (*Megachasma pelagios*, family Megachasmidae) from the Hawaiian Islands. In: *Proceedings of the Californian Acadacemy of Sciences* 43A (1983), pp. 87–110. This species together with the coelacanth is featured as a "monstrosity" on http://www.realmonstrosities.com/2013/06/coelacanth.html and http://www.realmonstrosities.com/2012/05/basking-shark.html (both accessed June 12, 2013).

Bibliography

Atz, James W.: The Lure of Latimeria. In: *Environmental Biology of Fishes* 61 (2001), pp. 347–351.

Balon, E. K. / M. N. Bruton / H. Fricke: A Fiftieth Anniversary Reflection on the Living Coelacanth, *Latimeria chalumnae*: Some New Interpretations of Its Natural History and Conservation Status. In: *Environmental Biology of Fishes* 23 (1988), pp. 241–280.

BBC Two: TV show *QI* (Quite Interesting), D: Ian Lorimer QI series H episode 3 Hoaxes – Is there such a thing as a fish-.mp4, 1:49); file viewable under http://www.youtube.com/watch?v=NdNEN4rUOTA (accessed November 11, 2013).

Berra, Tim M.: Some 20th Century Fish Discoveries. In: *Environmental Biology of Fishes* 50 (1997), pp. 1–12.

Bruton, Michael N. / Sheila E. Coutouvidis / Jean Pote: Bibliography of the Living Coelacanth *Latimeria chalumnae*, with Comments on Publication Trends. In: John A. Musick / Michael N. Bruton / Eugene K. Balon (eds): *The Biology of Latimeria Chalumnae and Evolution of Coelacanths*. Dordrecht / Boston: Kluwer Academic Publishers 1991, pp. 403–433.

Canadian Museum of Nature. http://www.nature.ca/discover/treasures/images/anim (accessed January 13, 2014).

Candolle, Alphonse de: Laws of Botanical Nomenclature Adopted by the International Botanical Congress Held at Paris in August 1867; Together with an Historical Introduction and Commentary by Alphonse de Candolle. In: *The American Journal of Science and Arts, Series II* 46 (1868), pp. 63–77 (First published in French, Genève / Bâle / Paris: H. Georg et J.-B. Bailliere 1867; Reprinted from the English translation published by L. Reeve and Co., London 1868).

Coelacanthwatch. http://www.coelacanthwatch.com (accessed June 5, 2013).

Chiquet, Simone: Was heisst eigentlich archivwürdig? In: *Schweizerische Zeitschrift für Geschichte = Revue suisse d'histoire = Rivista storica svizzera* 51,4 (2001): Überlieferungsbildung und Bewertung = Evaluation et formation des sources archivistiques, pp. 470–486.

Cook, Terry: From Information to Knowledge: An Intellectual Paradigm for Archives. In: *Archivaria* 19 (1984), pp. 28–49.

Council on Library and Information Resources: http://www.clir.org/pubs/reports/pub89/contents.html (accessed August 20, 2013).

De Sylva, D. P.: Mystery of the Silver Coelacanth. In: *Sea Frontiers* 12 (1966), pp. 172–175.

Derrida, Jacques: *Archive Fever. A Freudian Impression*. Chicago: University of Chicago Press 2008.

Dinofish. http://www.dinofish.com (accessed June 19, 2012).

Erdmann, Mark V. / L. Caldwell / M. Kasim Moosa: Indonesian 'King of the Sea' Discovered. In: *Nature* 395 (1998), p. 335.

Evenhuis, Neal L.: A Compendium of Zoological Type Nomenclature: a Reference Source. Electronic document, 2008. http://hbs.bishopmuseum.org/publications/pdf/bm-tp41.pdf (accessed September 18, 2013).

Fishbase. http://fishbase.org (accessed June 3, 2013)

Forey, Peter L. / Richard Cloutier: Literature Relating to Fossil Coelacanths. In: John A. Musick / Michael N. Bruton / Eugene K. Balon (eds): *The Biology of Latimeria Chalumnae and Evolution of Coelacanths*. Dordrecht / Boston: Kluwer Academic Publishers 1991, pp. 391–401.

Fossilmuseum. http://www.fossilmuseum.net (accessed January 17, 2014).

Foucault, Michel: Truth and Power. In: Id.: *The Foucault Reader. An Introduction to Foucault's Thought*, ed. by Paul Rabinow. London: Penguin Books 1991, pp. 51–75.

—: Nietzsche, Genealogy, History. In: Id.: *The Foucault Reader. An Introduction to Foucault's Thought*, ed. by Paul Rabinow. London: Penguin Books 1991, pp. 76–100.

—: *The Order of Things*. London / New York: Routledge 2002.

Fricke, Hans / Raphael Plante: Silver Coelacanths from Spain are not Proofs of Pre-scientific Discovery. *Environmental Biology of Fishes* 61 (2001), pp. 461–463.

Gilliland-Swetland, Anne: Enduring Paradigm, New Opportunities. The Value of the Archival Perspective in the Digital Environment. Washington, D. C.: Council on Library and Information Resources 2000. http://www.clir.org/pubs/reports/pub89/contents.html (accessed August 20, 2013).

Haworth, Penny (ed.): *Building on the South African Coelacanth Legacy. African Coelacanth Ecosystem Programme 2001–2013. 75th Anniversary 1938–2013*. Grahamstown: The South African Institute for Aquatic Biodiversity / The African Coelacanth Ecosystem Programme 2013. http://www.saiab.ac.za/uploads/files/saiab_acep_report_2013_website.pdf (accessed November 19, 2013).

Hissmann, Karen / Hans Fricke: Tauchfahrt zu einem Methusalem der Meere. In: *forschung – Das Magazin der Deutschen Forschungsgemeinschaft* 4 (2006), pp. 16–21 http://www.dfg.de/download/pdf/dfg_magazin/wissenschaft_oeffentlichkeit/forschung_magazin/2006/forschung_2006_4.pdf (accessed January 17, 2014).

Jenkinson, Hilary: Reflections of an Archivist. In: *Contemporary Review* 165 (1944), pp. 355–361.

Jolinck, Heyndrick Dirrecksen: A: Het Journaal van H. D. Jolinck. In: Jacob Cornelisz van Neck / Wijbrand van Warwijck / Johannes Keuning (eds): *De Tweede Schipvaart der Nederlanders naar Oost-Indië onder Jacob Cornelisz. van Neck en Wybrandt Warwijck 1598–1600. Journalen, Documenten en andere Bescheiden uitgegeven en toegericht door J. Keuning. Deel V, Eerste Stuk: De Reis naar Ternate*. 's-Gravenhage, Martinus Nijhoff 1947 (Werken uitgegeven door de Linschoten-vereeniging 50).

Musick, John A. / Michael N. Bruton / Eugene K. Balon (eds): *The Biology of Latimeria Chalumnae and Evolution of Coelacanths*. Dordrecht / Boston: Kluwer Academic Publishers 1991.

Nulens, Rik / Lucy Scott / Marc Herbin: *An Updated Inventory of All Known Specimens of the Coelacanth, Latimeria spp.* Grahamstown: South African Institute for Aquatic Biodiversity / African Coelacanth Ecosystem Programme 2011 (= Smithiana Special Publication, vol. 3) (http://www.saiab.ac.za/uploads/files/spec._pub._3_-_coelacanth_inventory.pdf, (accessed December 13, 2013).

Pouyaud, L. / S. Wirjoatmodjo / I. Rachmatika / A. Tjakrawidjaja / R. Hadiaty / W. Hadie: Une nouvelle espèce de coelacanthe: preuves génétiques et morphologiques: A New Species of Coelacanth. In: *Comptes Rendus de l'Académie des Sciences, Series III – Sciences de la Vie* 322 (1999), pp. 261–267.

Realmonstrosities. http://www.realmonstrosities.com (accessed June 12, 2013).

Seier, Andrea: Kategorien der Entzifferung. Macht und Diskurs als Analyseraster. In: Hannelore Bublitz / Andrea D. Bührmann / Christine Hanke / Andrea Seier (eds): *Das Wuchern der Diskurse. Perspektiven der Diskursanalyse Foucaults.* Frankfurt am Main: Campus 1999, pp. 75–86.

Smith, James L. Brierley: A Living Fish of Mesozoic Type. In: *Nature* 143 (1939), pp. 455–456.

—: The Living Cœlacanthid Fish from South Africa. In: *Nature* 143 (1939). pp. 748–750.

—: The Second Coelacanth. *Nature* 171 (1953), pp. 99–101.

Smith, M. M.: Latimeriidae. In: Id. / P. C. Heemstra (eds): *Smiths' Sea Fishes.* Berlin / Heidelberg: Springer 1986, pp. 152–153.

South African History Online. http://www.sahistory.org.za/dated-event/fishing-trawler-south-african-coast-catches-coelacanth (accessed June 3, 2013).

South African Institute of Aquatic Biodiversity. http://www.saiab.ac.za/home.htm (accessed November 19, 2013).

Taylor, Leighton R. / L. J. V. Compagno / Paul J. Strutsaker: Megamouth – A New Species, Genus, and Family of Lamnoid Shark (*Megachasma pelagios*, family Megachasmidae) from the Hawaiian Islands. In: *Proceedings of the Californian Acadacemy of Sciences* 43A (1983), pp. 87–110.

Thomson, Keith Stewart: *Living Fossil. The Story of the Coelacanth.* New York: Norton 1991.

Weinberg, Samantha / Andrea Stumpf / Gabriele Werbeck: *Der Quastenflosser. Die abenteuerliche Geschichte der Entdeckung eines seit siebzig Millionen Jahren vermeintlich ausgestorbenen Tieres.* Berlin: Argon 1999.

White, Hayden: The Narrativization of Real Events. In: *Critical Inquiry* 7 (1981), pp. 793–798.

—: The Politics of Historical Interpretation. Discipline and De-Sublimation. In: *Critical Inquiry* 9 (1982), pp. 113–137.

—: The Question of Narrative in Contemporary Historical Theory. In: *History and Theory* 23 (1984), pp. 1–33.

—: *Tropics of Discourse. Essays in Cultural Criticism.* Baltimore: Johns Hopkins University Press 1992.

Table of Figures

Fig. 1 © Virtual Fossil Museum. http://www.fossilmuseum.net (accessed January 17, 2014). Also available from http://upload.wikimedia.org/wikipedia/commons/6/67/Marjorie_Courtenay-Latimer_and_Coelacanth.jpg (accessed May 3, 2013).

Fig. 2 http://upload.wikimedia.org/wikipedia/commons/f/fa/Latimeria_Chalumnae_-_Coelacanth_-_NHMW.jpg (accessed May 3, 2013).

Fig. 3 © Aquamarine Fukushima. Also available from http://www.dinofish.com/cimages/Junkfood.jpg (accessed November 18, 2013)

Fig. 4 Hans Fricke / Raphael Plante: Silver Coelacanths from Spain are not Proofs of Pre-scientific Discovery. In: *Environmental Biology of Fishes* 61 (2001), pp. 461–463, here p. 462.

Fig. 5 Heyndrick Dirrecksen Jolinck: A: Het Journaal van H. D. Jolinck . In: Jacob Cornelisz van Neck / Wijbrand van Warwijck / Johannes Keuning (eds): *De Tweede Schipvaart der Nederlanders naar Oost-Indië onder Jacob Cornelisz. van Neck en Wybrandt Warwijck 1598–1600. Journalen, Documenten en andere Bescheiden uitgegeven en toegericht door J. Keuning. Deel V, Eerste Stuk: De Reis naar Ternate.* 's-Gravenhage: Martinus Nijhoff 1947 (Werken uitgegeven door de Linschoten-vereeniging 50), p. 56.

Fig. 6 © Canadian Museum of Nature. Also available from e. g. http:// www.nature.ca/discover/treasures/images/anim/trssketch-m.gif (accessed January 13, 2014); http://www.dinofish.com/cimages/csketch.jpg (accessed November 18, 2013).

Fig. 7 © Georg Berkemer.

The author wishes to express his particular gratitude to the Aquamarine Fukushima (Iwaki, Fukushima, Japan) and the Canadian Museum of Nature (Ottawa, Ontario, Canada) who kindly provided high resolution scans of their images.

Life on the Water

Soldiers, Slaves and Other Subalterns on Board of Ships in the Indian Ocean

Michael Mann

Introduction

The following text draws an outline of the social history of subalterns' lives on board ships in the western Indian Ocean between the eighteenth and the twentieth century. So far, many accounts have been written on the merchandise shipped between Europe and South Asia. Trade and commerce have dominated the maritime history. While plenty of eminent works have appeared in the last fifty years dealing with the social history of seamen, soldiers, convicts, servants and slaves transported across the Atlantic, the social history of the Indian Ocean has just recently become a focus of attention. The history of the life and the living conditions aboard ships is more or less untold—scattered across chapters of various edited volumes and articles in journals.[1] In many cases these smatterings of social history provide only limited background information. The widely untold history still lacks a systematic and synthesizing approach. This article attempts to summarise the current state of research literature.

1 The 'thickest description' of the life aboard ships in the nineteenth century is provided by Clare Anderson: 'The Ferringees are flying – the ship is ours!'. The Convict Middle Passage in Colonial South and Southeast Asia, 1790–1860. In: *Indian Economic and Social History Review* 42,2 (2005), pp. 143–186. Yet, see the recent publication by Roland Wenzelhuemer and Michael Offermann: Ships Newspapers and Passenger Life Aboard Transoceanic Steamships in the Late Nineteenth Century In: *Transcultural Studies* 1 (2012), pp. 77–121. http://archiv.ub.uni-heidelberg.de/ojs/index.php/transcultural/article/view/9363/3246 (accessed October 30, 2013).

Living conditions of a ship's subaltern crew and 'passengers' like sailors, slaves, soldiers, convicts, and kulis did not differ much until the middle of the twentieth century. It will be part of this article to highlight these similarities. The parameters of analysis will be 'race', gender, and class thus cutting across the categories of the labouring groups as fixed identities. The major part of this chapter, however, will describe the live and the living conditions of the said social groups according to thematic categories like "Fun and Fear", "Food and Drink", "Life and Limb" and "Disease and Death" thereby showing the multiple facets of life on the water. Generally seen the maritime social life differed much from life ashore as it was influenced by the captain's autocratic regimen and the limited space aboard. Yet, to some extent it may be comparable to a specific situation on shore—to the restricted life in jails with its inmates under almost permanent surveillance and control.[2]

The results of this outline will contribute to a global maritime social history. However they will also mark differences between a well-established Atlantic maritime history and the rather neglected maritime social history of the Indian Ocean. On the one hand, the maritime history of the Atlantic has been dominated by the discourse on emigrants—either those from various European countries and regions travelling West to a promised land or the discourse on the 'Black Atlantic' when Africans were shipped to the 'West' as well, however in that case not to the "land of the free". On the other hand, the maritime social history of the Indian Ocean has no specific focus at all. Contributions describe the life of convicts and kulis or lascars, seldom including crews, slaves and soldiers. Its history has been dominated by goods and wares, commerce and trade—people seemingly being of secondary importance. This article aims to change this biased perception and to open up the field or rather the Indian Ocean for a different approach which, with respect to humans and their social history, is not yet dominated by a specific historical narrative.

Precondition to the life on the water is the space provided by a ship's hull. The size of ships demonstrates how limited the space on board a ship was and how that space determined the life and the living

2 B. R. Burg: *Sodomy and the Pirate Tradition. English Sea Rovers in the Seventeenth-Century Caribbean.* New York: New York University Press 1995, pp. 107–110.

conditions aboard a vessel. In the seventeenth and eighteenth centuries an average ship sailing from Europe to the East Indies weighed 400 tons. Its built size was approximately 33 meters long, 10 meters large and 4 meters deep. Most English ships had less than 500 tons because ships over 499 tons had to take a priest on board, which caused additional costs, occupied space and did not fit in the captain's understanding of his sole authority on board a vessel. However, in the eighteenth century the tonnage of most of the ships employed in the East India trade ranged between 500 and 700 tons. In the 1760s, several ships were built with a tonnage up to 800 tons and in 1769 the *Princess Royal* came close to 900 tons. In 1787, the *Ceres* was put into service, the first one-thousand ton ship. In comparison, the tonnage of most ships operating in the trans-Atlantic trade varied between 150 and 200 tons.[3]

Crews and Passengers

The number of the crew on ships also varied between the Atlantic and the Indic. By the end of the seventeenth century the ratio between tonnage and men on trans-Atlantic ships was between 8.7 and 9.8 men per ton, in 1736 numbers ranged between 9.7 and 11 men per ton. On the European Atlantic coast ships were manned with a ratio of 6.7 and 7.7 tons per man. Ships bound to the Levant had a ratio of fifteen men per hundred tons. The high number of crew was required to serve the heavily armed ships of more than 200 tons equipped with 16 guns and 32 men at least.[4] The famous East-India-Men above 500 tons were as big and as armed as the contemporary Men-of-War.[5] Like the ships bound to the Mediterranean the East India bound vessels had to defend themselves against corsairs and pirates and, in times of European wars of which there were many particularly during the eighteenth century, against the trading company of a political enemy. In particular the English East India Company and the French Compagnie des Indes

3 Numbers for the years 1790–1791 show that out of a total of 1,156 ships built in England only 150 had more than 200 tons, Ralph Davis: *The Rise of the English Shipping Industry during the Seventeenth and Eighteenth Centuries*. Townbridge / London: Abbot 1972, pp. 70, 78, 261–262.

4 Ibid., pp. 58–59.

5 Jean Sutton: *The East India Company's Maritime Service, 1746–1834. Masters of the Sea*. Woodbridge: The Boydell Press 2010, p. 119.

Orientales did not like to be directly involved in the political quarrels and military conflicts of their governments at home.[6]

East India Bound ships of 500 tons were manned with approximately one-hundred sailors and officers, i. e. 20 men per ton. To get an impression about the composition of the crew, the example of the *Colchester*, a ship of 450 tons which sailed to India in 1703, may suffice. In addition to the master the ship had five mates and three midshipmen. The boatswain had two mates or servants, the gunner three and the carpenter four. There were a cook and a cooper, each had a mate, besides a steward and the captain's steward with two assistants, a purser, a caulker and his mate, a joiner and two tailors. There was also a surgeon with two mates. The rest of the crew consisted of 51 men and boys. Altogether the ship carried 89 men.[7] Apart from the officers the rest of the crew slept below the deck. The space between the decks varied from 4 feet 6 inches on small ships and up to 6 feet 6 inches on the largest ones. Yet most decks had a distance of 5 feet and 5 feet 6 inches. In addition to the members of the crew there was a varying number of soldiers, slaves, convicts and ordinary passengers. Depending on the size of the ship and the merchandise on board the vessel the number of subalterns, passengers of various kinds and officers could range from 50 up to several hundred persons. Yet, the ship's space not only was occupied by humans but also by animals like cattle, sheep, goats, chickens and pigs brought on as stock to be slaughtered from time to time.[8]

Generally, the life of passengers, soldiers or slaves, crew and officers was restricted to the limited space of a ship's hull. The voyage to the East Indies on average lasted about half a year. It is easy to understand that such voyages must have been extremely gruelling. The confinement on the ship tired the patience, nerves and health of everybody from the captain to the cabin boy, from the sailor to the soldier. The

6 *Vide* Michael Mann: Der ungeliebte Krieg: Compagnie des Indes und East India Company als Kombattanten in einem globalen Konflikt, 1742–1763. In: Sven Externbrink (ed.): *Der Siebenjährige Krieg als (europäischer) Weltkrieg*. Berlin: Akademie Verlag 2010, pp. 99–126.

7 Davis: *The Rise of the English Shipping Industry*, pp. 110–111.

8 Marcus Rediker: *Between the Devil and the Deep Blue Sea. Merchant Seamen, Pirates and the Anglo-American Maritime World, 1700–1750*. 12th ed. Cambridge: Cambridge University Press 1987 [2007], pp. 160–161.

size of a ship as well as the length of the voyage influenced the life and the number of conflicts. "[T]he close and isolated quarters of the ship with its food shortages, monotony and frustration of confinement bred conflict and animosity like a hot-house."[9] Tension and stress between the officers and the crew or between any of the groups on board could rise to explosive proportions. Most tensions existed between the officers who were often drunk, and the common sailor. The despotic and often brutal regime organised in a strict hierarchical order with highest authority of the captain was most likely the reasons why the solidarity among the members of the crew was fairly high.[10]

On the one hand living conditions strengthened the solidarity among the members of the crew, on the other hand, they contributed to the fragility of a ship's temporal society or aggressiveness among the passengers and the rest of the ship's population. The confinement was only manageable by the strict command of the captain, that often turned into a despotic regimen of control, discipline and punishment on a limited space with no escape. Extremely limited space, restricted mobility, sensory stimulation, leisure and social activities, food, and, of course, hard work determined the everyday life and culture aboard a ship. As Michel Foucault noted, ships are "Heterotopias par excellence [...] the ship is a piece of floating space, a placeless place, that lives by its own devices, that is self-enclosed and, at the same time, delivered over to the boundless expanse of the ocean."[11] Or, as Rediker summarises it: "There was too little space aboard the ship and too much outside."[12]

Life was not only determined by the cramped situation which forced people from different social backgrounds to live together for months but also by the intensity of life and death. A high turnover of a ship's personnel and passengers, a high amount of sickness and thus a high mortality rate was typical for the maritime culture in the 'Age of Sail'. Within the course of a single day seamen or slaves could be punished

9 Ibid., pp. 39–40.

10 Karel Degryse: Social Conditions and Tensions on Board the Eighteenth Century East India Ships. In: K. S. Mathew (ed.): *Mariners, Merchants and Oceans. Studies in Maritime History*. Delhi: Manohar 1995, pp. 341–346, esp. 342–344.

11 Michel Foucault: Different Spaces. In: Id. / James D. Faubion: *Aesthetics, Method and Epistemology*. New York: The New Press 1998, pp. 175–185, esp. pp. 184–185.

12 Rediker: *Between the Devil and the Deep Blue Sea*, p. 159.

severely even for minor misdeeds. Often punishment had to be watched by the passengers, the crew and the slaves, for the latter two groups as a means of deterrence, for the first group as an example of the deterring justice executed by the captain.[13] Life aboard a ship was dense and intense. The very same day could see the birth of a child, the baptism of another, the death of a passenger, the punishment of a crime, or the suicide, by jumping over board, of a slave or crew member.[14]

Depending on the creed, religious service and religious faith could vary extremely. A daily Mass was held for all people aboard Portuguese ships. People lived in a certain religious atmosphere often praying for the help of god for a safe journey. Mass, litanies, a *Te Deum*, confessions, prayers completed the religious services administered by the priest.[15] On Portuguese ships religious rites were practiced rather regularly or often, whereas on English ships of the early eighteenth century creed, belief and religious service were of less importance. It is reported that a priest complained about the swearing and cursing of crew members during a tempest. When the crew finally started praying the captain replied "Oh, I am afraid that if they have stopped swearing and started praying there is no hope for us." If the wind blew strong regular prayers took place once a week, if there was little wind, no prayer took place at all. Even contemporaries like Daniel Defoe noticed that there was a process of de-christianisation going on at least among the members of the crew who actively discouraged or even eradicated religious belief.[16]

The crew and passengers on Arabian, African and Indian ships performed specific rituals. An ancient Sanskrit text of the tenth or eleventh century mentions the offering of holy water, curd, rice, milk and garland to the sea. A Persian document from the eighteenth century describes a ritual according to which a ship's crew assembled on the eve of departure throwing coconuts into the water if no favourable

13 *Vide* the passenger giving his witness account in court in Davis: *The Rise of the English Shipping Industry*, p. 154.

14 Sutton: *The East India Company's Maritime Service, 1746–1834*, pp. 179, 239–240.

15 Maria de Jesus Teodoro Dos Martires Lopes: Seamen, Soldiers and Crew of the *Carreiro da India* in the Second Half of the Eighteenth Century. In: K. S. Mathew / Satish Chandra (eds): *Ship-building and Navigation in the Indian Ocean Region, AD 1400–1800*. New Delhi: Munshiram Manoharlal Publishers 1997, pp. 132–139, esp. pp. 137–138.

16 Rediker: *Between the Devil and the Deep Blue Sea*, pp. 169–175, quote p. 169.

wind or calm was to be expected. Another maritime superstition existed with respect to human bones or corpses aboard a ship, crew and passengers fearing that the ship might be overtaken by a tempest or calm.[17] If tempests were feared because of the power exposed by nature daring to sink a ship, calm was likewise feared because of the heat. A European eyewitness of the seventeenth century vividly described the situation on board a ship during calm weather:

> This induced the passengers and rich Moslem merchants to try all kinds of sorcery and witchcraft to raise the wind. Sometimes they tied to the mast little paper flags inscribed with the praises of their prophet Muhammad. Sometimes they carried large basins all over the ship to the principal merchants and the whole crew, just as we make a collection in France during the sermon or in a High Mass, in aid of the poor. But here the basins were filled with rice, meat, sweets, butter, fruit and other food, which were to be thrown into the sea all around the vessel. They also obliged all the followers of their faith, headed by their priests, to bathe their bodies in the sea, in order to wash away the dirty impurities they commit with their young slaves, of which there were more than 200 in the ship. Again they searched all the chests and baggage of the Persian and Armenian passengers, saying they sometimes bring back from India the bones of relations which they want to bury in their own country.[18]

On Indian Ocean ships seamen, slaves and passengers, mostly traders sometimes accompanied by their wives, experienced the same fears as did sailors and passengers, soldiers and slaves on European ships. European Ships were baptised to detain bad luck and many Arabian ships were painted with the blood of a goat to revive the dead wood used for the construction of the vessel and in this way protecting the ship from bad luck. For the same reason the keel of Arabian ships under construction was painted with a goat's blood. A report from 1870 states that when launching a ship at the western coast of India, an odd number of goats was sprinkled with water, their heads were rubbed with cinnamon, they were beheaded and the heads fixed to the ship's bow. The transom stern of a ship was painted in bright colours

17 A. Jan Qaisar, From Port to Port: Life on Indian Ships in the Sixteenth and Seventeenth Centuries. In: Ashin Das Gupta / M. N. Pearson (eds): *India and the Indian Ocean 1500–1800*. Calcutta / New York: Oxford University Press 1987, pp. 331–349, cf. pp. 334–335.

18 *The Travels of Abbé Carré in India and the Near East, 1672–74*, quoted in Qaisar: From Port to Port, p. 343. Loss of life due to the heat could be fairly high, *vide* Johann Christian Hoffman: *Ostindische Voyage 1671–1676* [1680], reprint ed. by Günter E. Th. Bezzenberger. Kassel: Thiele & Schwarz 1983, pp. 21–23.

depicting floral, animal, geometrical, and figural ornaments thus drawing a connection between the material and the magic elements a ship consisted of.[19]

From the eighteenth century onwards the increasing number of European ships sailing into and across the Indian Ocean joined a long established coastal and inter-regional and even inter-continental maritime trade relying on *dhaus*. Actually the term *dhau* is a Swahili word not used by Arabs who rather named their ships, according to the construction of the hull, *baghadah*, *ganja*, *sambūq*, *jihāzi* etc.[20] However, as the British increasingly dominating the Indian Ocean maritime trade in the nineteenth century they popularised the term *dhau*.[21] It comprised a wide range of vessel types ranging from 60 to 100 tons. Smaller *dhaus* of only 10 to 40 tons dominated the trade along the eastern African coast from Mozambique to the Arabian Peninsula and the Persian Gulf stretching to the western coast of the Indian peninsula.[22]

Dhaus were fairly manoeuvreable ships, most of them not longer than 15 metres and even in the 'Age of Steam' and iron hulls they were pure wooden constructions. Yet, shipbuilders around the Persian Gulf also constructed the majestic *baghlas*, large ships of up to 600 tons and especially designed for the deep water trade of the Indian Ocean. On medium sized *dhaus* the average number of seamen was 15 to 20. Often the members of a crew did not only originate from the same region or coastal city but belonged to the same clan or even family. Ships sometimes also carried 80 to 120 *hajj*-pilgrims to the Red Sea. At times ships were overloaded with pilgrims, men, women, children and sick people as well as animals and trading goods leading to waves coming overboard drenching all of them. British documents from the

19 Erno Wiebeck / Hermann Winkler: *Segler im Monsun. Die Dau am Indischen Ozean.* Rostock: Neuer Hochschul-Schriften Verlag 2000, pp. 51–55.

20 Dionisius A. Agius: *In the Wake of the Dhau.* Reading: Ithaka 2002, pp. 36–48. Wiebeck / Winkler: *Segler im Monsun*, pp. 11–20.

21 Ibid., pp. 7–8; Agius: *In the Wake of the Dhau,* pp. 33–36.

22 In fact *dhaus* were built and developed as ships from ancient times diversifying into a multitude of different types of ships. For the variety of wooden ships in the Gulf region and around the Arabian Peninsula *vide* Agius: *In the Wake of the Dhau* pp. 49–53. For the east African and western Indian ship building industry see Wiebeck / Winkler: *Segler im Monsun*, pp. 57–104. Additionally cf. George F. Hourani: *Arab Seafaring in the Indian Ocean in Ancient and Medieval Times.* Princeton, NJ: Princeton University Press 1995, pp. 87–128.

beginning of the eighteenth century testify that Arabian and Persian vessels (*ghurabs*) in fact did carry between 14 and 20 guns hence contradicting the European stereotype that the empires of the East did not have a navy because their ships were inable to carry heavy guns.[23]
The average journey across the Arabian Sea took up to 30 days, along the western coast of South Asia up to 22 days. After the Suez Canal had been opened in 1869 shortening the passage from and to Europe from six months to six weeks, *dhaus* nevertheless survived in the western Indian Ocean dominating the coastal trade whilst the deep water trade became the domain of the large European shipping companies, in particular P&O (Peninsular and Oriental Steam Navigation Company) and the British India Steam Navigation Company. However, due to construction and energy costs Europeans kept on constructing conventional sailing ships right into the first half of the twentieth century, in particular for transporting bulky wares from Asia to Europe as well as shipping Indian indentured labourers across the Ocean.[24] Arab, African and South Asian ships specialised on a coastal port-to-port trade including the transport of slaves, ivory, rice and spices. It was only at the beginning of the twentieth century that the *dhau* trade and traffic diminished considerably.[25]
Wooden ships always had the problem of water leaking into the hull of the ship that fell into the hold or bilge. This was a particular problem of hulls with more than one deck since the collecting water rotted the wood. As antiseptics and chemicals to protect the wooden planks were unknown in these days the smell of bilge-water was considered by the seamen to be part of a seaman's business. According to a late eyewitness everyone who complained about the revolting stench was regarded as effeminate: 'borne ashore' was the usual sneer.[26] A similarly

23 Wiebeck / Winkler: *Segler im Monsun*, pp. 41–42, 49; Agius: *In the Wake of the Dhau*, p. 15, fn. 22.

24 Ravi Ahuja: Lateinsegel und Dampfturbinen. Der Schiffsverkehr des Indischen Ozeans im Zeitalter des Imperialismus. In: Dietmar Rothermund / Susanne Weigelin-Schwiedrzig (eds): *Der Indische Ozean. Das afro-asiatische Mittelmeer als Kultur- und Wirtschaftsraum*. Wien: Promedia 2004, pp. 207–226.

25 During the 1870s some 43,000 *dhaus* were sailing in the western Indian Ocean, in the middle of the twentieth century their number had fallen to 5,300, see Wiebeck / Winkler: *Segler im Monsun*, pp. 55–56.

26 William E. Home: *Merchant Seamen. Their Diseases and their Welfare Needs*. London: Murray 1922, p. 16.

revolting stench originated from the forecastle, the quarters of the sailors situated before the mast, whether wooden or iron hull. At the end of the nineteenth century, the seamen's space on an average iron ship was 846 cubic feet or even less, iron wall outside, iron partitions inside. Twelve beds sometimes made out of wood arranged in pairs and threes left and right. Under the berths seamen kept their belongings, private items in their beds. Seamen's food was kept in special lockers with holes pierced in them for ventilation.

As in wooden ships the outer hull of an iron ship was not watertight thus water leaked into the ship. The dampness of the deck was increased by the condensation and moisture from the breath of the men in cold weather, the water collecting at the ceilings, running down the walls, dropping on the sailors' beddings and bodies. Worse than the stuffiness, was the moistness aboard a ship. Altogether an iron ship was

> [...] ill ventilated, the smell of the cable locker is in it, the smells of the paint-room, perhaps of the offensive cargo the ship is carrying; it is the only smoking-place out of the weather that the crew have, they have only buckets to wash in, and their beddings and cloths are dirty too, for they have no seats but the shelves mentioned, so when they are tired they lie down at once down on their beds in their boots, firemen off watch in their engine-room clothes; and, by the bye, the privy is closely adjacent and likely adds to its aroma.[27]

It seems quite clear that the ensuing 'Age of Steam' did not necessarily mean an improvement with respect to the living conditions of sailors and seamen in comparison to the 'Age of Sail'. Stuffiness, wetness, tightness and fetidness remained the sailors' environment on board a ship, in many instances until the middle of the twentieth century.

Sailors, Seamen and Lascars

The tension between the officers and the common sailor resonates from the very strong class segregation and the prejudices. According to seventeenth and eighteenth centuries' judicial records, sailors represented the 'scum of the society'. Generally the members of a ship's crew were recruited in the port towns of the great seafaring countries of north-western Europe. Often they were the children of labourers, farmhands and poor peasants and seamen.[28] Highly interesting is

27 Home: *Merchant Seamen*, pp. 21–23.

28 Davis: *The Rise of the English Shipping Industry*, p. 114

the account of a contemporary Dutchmen regarding the social and geographical background of people seeking employment with the Vereenigde Oostindische Compagnie (VOC). In his *Oost-Indise Spiegel* Nicolaas de Graaff, a ship surgeon, wrote at the end of the seventeenth century, that the VOC was

> […] een grote toevlugt voor alderhande vreemdelinge en uitheemse natie, als Polakke, Sweeden, Deenen, Noor-luyde, Jutte, Hamborgers, Bremers, Lubekkers, Dantsikers, Konixbergers, Hoogduytse, Oosterlingers, Westfaalders, Bergse, Gulikse, Kleefse, en voort alderhande Moffen, Poepe, Knoete, Hannekemaijers, en andere groene kassoepers, die't gras nog tussen de tanden steekt […].[29]

Looking at the geographical background of the people seeking employment with the VOC, it is evident that the vast majority came from middle and eastern European countries and cities. The recruits of the crew were fairly young. For the Anglo-American Atlantic numbers indicate that two-thirds of the crew members were in their teens and twenties.[30] It is interesting to note that many Germans were recruited as soldiers or crew members, seldom as writers or merchants, posts which remained reserved for Dutch people.[31]

Regarding the East India Company (EIC), the composition of a ship's crew was even worse. From the last quarter of the eighteenth century onwards, the Admiralty had enforced by Parliament that the press gangs of the EIC were to recruit seamen in the port towns of Great Britain only after the recruiters of the Royal Navy had scoured for the sailors. They skimmed off the healthier and stronger sailors from the vast amount of seamen seeking employment. Accordingly, the physical state and the health condition of the EIC seamen was worse than that of a sailor on a warship deeply influencing the state of health and

29 Nicolaas de Graaff: Oost-Indise Spiegel (1703). In: J. C. M. Warnsinck (ed.): *Reisen van Nicolaus de Graaff. Gedaan naar alle gewesten des Werelds beginnende 1639 tot 1687 incluis.* 's-Gravenhage: Nijhoff 1930 [1703], p. 47. "[…] a large refuge for many foreigners from strange nations like people from Poland, Sweden, Denmark, Norway, Jutland, Hamburg, Bremen, Lubeck, Danzig, Koenigsberg, High-Germans (Southern Germany), Eaterners, Westfalen, Berg, Gulikse, Kleeve, and all sorts of rough guys, bad characters, scallywags, and other green grasshoppers which still have grass between their teeth."

30 Rediker: *Between the Devil and the Deep Blue Sea*, p. 299.

31 Gita Dharampal-Frick: *Indien im Spiegel deutscher Quellen (1500–1750) Studien zu einer interkulturellen Konstellation.* Tübingen: Niemeyer 1994, p. 58.

the rate of mortality during the long lasting voyage.[32] With respect to sailors' living conditions aboard a ship in the eighteenth century, a certain Dr Johnson remarked in 1759: "No man will be a sailor who has contrivance enough to get himself into a ship; for being in a ship is being in a jail with the chance of being drowned […] A man in a jail has more room, better food, and commonly better company." In how far the company is comparable remains unclear and a crew's food was probably better during the first stage of a journey.[33]

From their earliest days in the Indian Ocean, European trading companies also used to employ locally recruited or available seamen on their "country ships", European or Asian built ships employed in the trade between the countries along the shores of the Indic.[34] Such sailors were by far cheaper and, at the same time, regarded fitter for the hard work on board a ship in the hot weather of the seas. In the eighteenth century the Persian term *las(h)kar* was used by the British for denominating sailors from South Asia as well as the non-combatant military personnel of a warship (e.g. gun-lascars). However, lascar was not the only word to name sailors in the Indian Ocean. *Mallah*, *jehazi* or *khalasi* were also denominations for native seamen and preferred by them as lascar, like "coolie", was regarded as a discriminatory term.[35] In the 'Age of Steam' the British employed lascars in a steadily increasing number on the ships of their merchant fleet.[36] In 1881, 16,673 of the

32 Michael Mann: *Bengalen im Umbruch. Die Herausbildung des britischen Kolonialstaates, 1754–1793.* Stuttgart: Steiner 2000, p. 110.

33 Quoted in Davis: *The Rise of the English Shipping Industry*, p. 154, cf. fn. 1.

34 The Portuguese already manned their ships with Asian, African and Arabian sailors, cf. G. Balachandran: Recruitment and Control of Indian Seamen: Calcutta, 1880–1935. In: *International Journal of Maritime History* 9,1 (1997), pp. 1–18, esp. p. 3.

35 Ravi Ahuja: Die 'Lenksamkeit' des 'Lascars'. Regulierungsszenarien eines transterritorialen Arbeitsmarktes in der ersten Hälfte des 20. Jahrhunderts. In: *Geschichte und Gesellschaft* 31,3 (2005), pp. 323–353, esp. p 346.

36 Jonathan Hyslop: Steamship Empire. Asian, African and British Sailors in the Merchant Marine, c. 1880–1945. In: *Journal of Asian and African Studies* 44,1 (2009), pp. 49–67; Ravi Ahuja: Subaltern Networks under British Imperialism. Exploring the Case of South Asian Maritime Labour (c. 1890–1947). In: Jan-Georg Deutsch and Brigitte Reinwald (eds): *Space on the Move. Transformations of the Indian Ocean Seaspace in the Nineteenth and Twentieth Century.* Berlin: Schwarz 2002, pp. 39–60, esp. pp. 39–40, 47–50. *Vide* also the revised version of the aforementioned article, id.: Networks of Subordination – Networks of the Subordinated: The Ordered Space of South Asian Maritime Labour in an Age of Imperialism (c. 1890–1947). In: Ashwaini Tambe / Herald Fischer-Tiné (eds): *The Limits of British Colonial Control in South Asia. Spaces of Disorder*

204,470 (8.2 per cent) seamen were said to be lascars. In 1914, the total number of seamen employed by the British merchant fleet had risen to 295,652, lascars comprising 51,616 persons (17.5 per cent). During the following decades the total number of sailors dropped whilst the proportionate number of lascars increased (1838: 192,375 sailors—50,700 lascars = 26.4 per cent).[37]

From the second half of the nineteenth century onwards a recruiting network for lascars developed comprising 'catchment areas' such as Sylhet District in eastern Bengal or the north-western part of the Panjab. Recruits were concentrated in Bombay and Calcutta and employed as sailors by shipping companies according to their skills, age, physical condition etc.[38] The Asiatic Articles put lascars under a strict working regime comparable to that of the indentured labourers (kulis). In a series of legislative acts starting in 1814, the Asiatic Articles regulated the contracts of lascars setting wages at one-fifth to one-third the wages of the European sailors.[39] Lascars were given a contract for a certain amount of time, one year, eighteen months or two years thus providing dependable labour supply for short and medium range. This was in contrast to the 'Age of Sail' when crews were recruited for one particular voyage. In any case the lascar had to return to his port of departure whether on the ship he came with or on another homebound vessel.[40] And finally the Asiatic Articles restricted the settlement of lascars in Britain.[41] Additionally, according to the Merchant Shipping Act of 1914 a lascar was supposed to require less space on board a ship than European seamen: 72 instead of 120 cubic

in the Indian Ocean Region. London / New York: Routledge 2009, pp. 13–48. From the same root the east African term *askari* is derived becoming the Swahili term for soldier, cf. Caroline Adams: *Across Seven Seas and Thirteen Rivers. Life Stories of Pioneer Sylheti Settlers in Britain*. London: THAP Books 1987, p. 15.

37 C. H. Dixon: Lascars. The Forgotten Seamen. In: Rosemary Ommer / Gerald Panting (eds): *Working Men who Got Wet*. St. Johns: Maritime History Group, Memorial University of Newfoundland 1980, pp. 263–281, here p. 281.

38 For the Calcutta recruitment system *vide* Balachandran: Recruitment and Control of Indian Seamen, pp. 1–18.

39 Ibid., p. 2.

40 Janet J. Ewald: Crossers of the Sea. Slaves, Freedmen, and other Migrants in the Northwestern Indian Ocean, c. 1750–1914. In: *American Historical Review* 105,1 (2000), pp. 69–91, esp. p 76.

41 Adams: *Across Seven Seas and Thirteen Rivers*, p. 16. The book is based on the personal memory of lascars immigrating into London.

feet per person.[42] Obviously the Asiatic Articles turned the lascar into a legal category separate and subaltern from the European sailor.[43] Most crews of sailing as well as steam ships were not of British origin. Members were of Chinese, Malay Archipelago, Indian, Arabian or African origin, some of them hired labourers, some employed with companies, some on non-European ships even slaves. Many lascars were slaves on board a ship trafficking slaves from the interior of Eastern Africa along the shores of the Arabian Sea and to the well-known port cities and slave markets.[44] In the nineteenth century, the boundaries between the categories of slave, freedman, seaman, lascar, wage-worker and indentured labourer seem to have blurred in many cases. Working on ships and in the port towns of Eastern Africa, Arabia and South Asia, they all played a vital role in building and sustaining the Indian Ocean maritime labour regime and maritime culture from the eighteenth century right into the middle of the twentieth century.[45]

Soldiers

From the middle of the eighteenth century, the EIC regularly sent recruits from Britain to India to facilitate its military operations. Together with the King's troops, they had successfully conquered large parts of the subcontinent by the end of the century. However, the EIC had to face continuous difficulties in procuring the necessary manpower since, in contrast to the regular army whose officers were free to roam the country for recruits, the EIC was forced to deal with agents who supplied the EIC with recruits for financial rewards. Rumours about the sanitary conditions aboard ships and tropical diseases in India reduced the willingness of recruits to enrol for the East India Service. In 1778 a regular regiment mutinied on the mere rumour of being sold to the EIC. Even newspapers mentioned the soldiers'

42 F.J.A. Broeze: The Muscles of Empire – Indian Seamen and the Raj, 1919–1939. In: *Indian Economic and Social History Review* 18,1 (1981), pp. 43–67, here p. 44, fn. 3. Other numbers indicate 36 cub feet and 72 cub feet resp., cf. Ahuja: Subaltern Networks under British Imperialism, p. 40.

43 Ravi Ahuja: Mobility and Containment. The Voyages of South Asian Seamen, c. 1900–1960. In: *International Review of Social History* 51 (2006), Supplement, pp. 111–141, esp. pp. 117–119.

44 Michael Mann: *Sahibs, Sklaven und Soldaten. Geschichte des Menschenhandels rund um den Indischen Ozean.* Darmstadt: WBG 2012, p. 194.

45 Ewald: Crossers of the Sea, p. 70.

situation stating that "[w]hen they are aboard, they are of no defence but a Weakness to the ship. A great proportion of them die in the passage [...] the rest carry insolence, mutiny, profligacy, debauchery, and disease into their Armies in India."[46]

During the 1780s the number of EIC recruits dropped continually, starting with 2,047 in 1771 and ending with 665 in 1780. Additionally, the quality of recruits decreased and an increasing number of them was below twenty years of age, and even below sixteen. During the Revolutionary Wars the EIC sent 3,817 men to India, in 1793 and in the following year another 1,877.[47] Yet the King's army in India was also short of men. Therefore, in 1782 some 2,800 soldiers were sent from Hanover to Madras to serve British colonial interests in South Asia.[48] After the EIC had officially asked for the supply with troops, George III, Prince Elector of Hanover and King of England, ordered two infantry regiments to be recruited in central Germany and sent to Madras; as he had done with about 19,000 'Hessians' recruited for the theatre of war in the British colonies in North America between 1776 and 1782.[49]

Of the 2,800 soldiers sent to South India less than half returned home after two years. Many of the so called 'Hessian' recruits came from present-day eastern parts of Lower Saxony, many more from the northern parts of Hessen and adjacent parts of Thuringia.[50] Since recruitment and transportation of the 'Hessians' sent to Madras in 1782 is fairly well documented, their life, sufferings and death when transported serves as an example and is detailed below. Apart from these heavy troop-transporters to South Asia, all European trading

46 Arthur N. Gilbert: Recruitment and Reform in the East India Company Army, 1760–1800. In: *Journal of British Studies* 15 (1975), pp. 89–111, esp. pp. 91–93, here p. 93.

47 Ibid., pp. 90, 103.

48 The recruitment and finances of these troops was highly debated between Parliament on the one hand and the Court of Directors of the East India Company on the other hand, cf. Mann: *Bengalen im Umbruch*, pp. 321–326. Also Gilbert: Recruitment and Reform in the East India Company Army, 1760–1800.

49 Rodney Atwood: *The Hessians. Mercenaries from Hessen-Kassel in the American Revolution*. Cambridge: Cambridge University Press 1980, *passim*. Stephan Huck: *Soldaten gegen Nordamerika. Lebenswelten Braunschweiger Subsidiärtruppen im Amerikanischen Unabhängigkeitskrieg*. München: Oldenbourg 2011.

50 Viktor von Diebitsch: Die kurhannoverschen Truppen in Ostindien 1782–1792. In: *Hannoversche Geschichtsblätter* 1 (1898), pp. 67–128, esp. pp. 74–76.

companies regularly transported smaller number of soldiers to their forts in South Asia and Eastern Africa. During European warfare sometimes one or two regiments were shipped as in the case of two regiments from Württemberg.[51]

Kulis

In the time between the abolition of the slave trade in 1807 and when the abolition of slavery in the British Empire had been resolved by Parliament in 1834, debates started on how to substitute the slave workforce. It had become clear that slaves would not be willing to work as wage labourers on the plantations of their former masters. The 1830s were a decade of experiments which involved the government in London, the governor general of British India, the governor of Mauritius and representatives of the Caribbean plantocracy. Finally, an alternative was found with the indentured labour system based on Indian kulis. Rules and regulations were resolved during the following decades and stipulated recruitment, transportation, medical care, sanitary conditions on board of ships as well as the duration of the contract. Between 1846 and 1932 more than 28 million kulis left India of whom some 26 million returned. Most of the kulis originated from remote hilly regions of eastern and southern India, some from upper India. The majority of kulis was *dalits* (low and out castes) and *adivasi* (tribals) yet we also find members of an impoverished agrarian strata as well as urban artisans and even Brahmans.

It is interesting to note that kulis did not simply replace slaves but that, at least in the western part of the Indian Ocean, wage, indentured, bonded and slave labour coexisted well into the twentieth century. Globalisation of the capitalist labour market also facilitated and promoted the integration of markets, specialisation of production, and the diversification of labour forces.[52] The kuli-recruitment system was

51 In 1787, Herzog Karl von Württemberg rented out an infantry regiment including artillery units altogether consisting of roughly 2,000 soldiers to the VOC. First, the so called „Kap-Regiment“ was stationed at the Cape of Good Hope afterwards shipped to Trincomalee on Ceylon and finally transferred to Indonesia. After 1807 no news are available. Only a few soldiers of that regiment came home after more than two decades, Hans-Martin Maurer: Das Württembergische Kapregiment. Söldner im Dienst früher Kolonialpolitik (1787–1808). In: *Zeitschrift für Württembergische Landesgeschichte* 47 (1988), pp. 291–307.

52 Mann: *Sahibs, Sklaven und Soldaten*, pp. 211–213.

organised partly by the state and partly by private enterprises. Shipping as well as the plantations were in private hands. Contracts regulated the period of work, lasting from three up to five years. Prolongation instead of returning home was often possible and in many cases preferred by the plantation owners since that reduced recruitment costs. Contracts also regulated the working conditions - particularly working hours - and restricted the freedom of movement. Part of their contract was the voyage to the overseas plantation.[53]

Slaves

Slavery and slave trade was an age-old business in the coastal areas of the Indian Ocean. Yet, in contrast to the trans-Atlantic slave trade organised by Europeans from the sixteenth century onwards and lasting until the end of the nineteenth century, the slave trade of the Indian Ocean took place on a much smaller scale with respect to people caught, sold and shipped to countries around the Indian Ocean.[54] Much has been written on the infamous Middle Passage of the Atlantic slave trade, the horrible conditions on board as well as the extremely high mortality rate. This continues a discourse constructed and established by contemporary late eighteenth and early nineteenth Evangelical Abolitionists highlighting the horrors of wretched slaves on board the ships dying by the thousands, and continued by the estimates of academic research at the closing decades of the twentieth century, maintaining that 9 to 12 million Africans were shipped across the Atlantic.[55]

The dimensions of this trade are far less dramatic in the Indian Ocean, where it was organised differently and part of a complementary trading pattern. Like ivory, slaves were luxury items and therefore in most cases not as crammed on board a ship as the well-known rawing of the trans-Atlantic slave trade ship in England suggests. According to

53 Michael Mann: Circulation and Migration. In: Philippa Levine / John Mariott (eds): *The Ashgate Research Companion to Modern Imperial Histories.* Farnam / Burlington: Ashgate 2012, pp. 377–393; id.: Migration – Re-migration – Circulation: South Asian Kulis in the Indian Ocean and Beyond, 1840–1940. In: Donna R. Gabaccia / Dirk Hoerder (eds): *Connecting Seas and Connected Ocean Rims. Indian, Atlantic, and Pacific Oceans and China Seas Migrations from the 1830s to the 1930s.* Leyden / Boston: Brill 2011, pp. 108–133.

54 Mann: *Sahibs, Sklaven und Soldaten*, pp. 124–149.

55 Herbert S. Klein: *The Atlantic Slave Trade.* 8th ed. Cambridge: Cambridge University Press 2005 [1999], pp. 130–160 *passim.*

reports of the nineteenth century the average number of slaves on board an Indian Ocean ship was between 5 and 20. They were part of composite wares transported along the coasts or across the western Indian Ocean. From time to time it is reported from Portuguese ships that they had some hundred slaves on board. Yet, this seems to have been the exception to the rule of small scale transports. After the abolition of slavery in the British Empire in 1834, slave trade and slavery did not disappear in the Indian Ocean. On the contrary, it increased considerably. The Portuguese shipped an increasing number of slaves from Mozambique to Brazil thus tightening the colonial connections of the southern Indic with the southern Atlantic whereas the establishment of the Merina Empire on Madagascar also caused an increase in bonded and slave labour from the 1820s onwards. The same is true for the new Zanzibar-Muscat Empire of eastern Africa and southern Arabia from the 1840s.[56]

From the middle of the seventeenth century Europeans became massively involved in the slave trade of the western Indian Ocean but, in stark contrast to the Atlantic Ocean, they did not at any time dominate it since Africans and Arabs as well as South Asians remained the major actors in that trade. So far academic research cannot provide numbers of people involved in the slave trade and the numbers of slaves actually captured, shipped and sold. Nevertheless it can be stated that the maritime slave trade off Eastern Africa up to the Arabian Peninsula and the Persian Gulf increased during the nineteenth century. More research is needed to investigate the nature and the dimensions of that trade.

Convicts

From the end of the eighteenth to the middle of the nineteenth century the EIC established several penal colonies for Indian convicts. Benkulu served as place of detention from 1787 to 1825, succeeded by Melaka and Singapore (1825-60), followed by the Andaman Islands (which had already been a penal colony from 1793 to 1796) from 1858 until the end of British colonial rule in India 1947. Every year as many as eighteen convict ships left British India and about the same number brought prisoners from the British colonies in Southeast Asia to British

56 Mann: *Sahibs, Sklaven und Soldaten*, pp. 73–76, 136–149, 180–181.

India. The number of convicts transported ranged from a few dozen up to more than 200. Altogether the EIC transported more than 2,000 convicts to Benkulu and at least 300 to the Andamans during its first period as a penal colony. A conservative estimate indicates that at least 20,000 convicts were transported to the Strait Settlements. Additionally, some 1,500 prisoners from Ceylon / Sri Lanka were transported to Southeast Asia whilst between 1,000 and 2,000 convicts were brought from there to British India. Convicts were transported on EIC ships as well as on private ships, particularly after the EIC had lost its trading monopoly in 1834.[57]

There were no specific rules for convicts regarding sanitary conditions, food rations, clothing etc. during transportation. After the attempted rebellion on the *Catherine*, rules stipulating the proportion between the crew and the convicts were set up. Three convicts to one European sailor, one convict to one Asian seaman. Commanders were reminded not to trust the convicts and to check their fetters and irons daily. Security rules seem to have been strict but, as evidence indicates, treated rather lax. In contrast to the slave ships of the Atlantic, the ships transporting convicts in the Indic were not particularly fitted for their transportation. Convicts were accommodated below deck, the space allotted to them measuring six by two feet—the same space which was allotted to lascar. However, Bengal ships only provided space of six by one foot. Scarcity also characterised food rations and medical care. To some extent this neglect of provisions, care and space turned into the weakness of a convict ship's regime and in some instances provided the reason for rebellion.[58]

Sex and Sensuality

Sex was not only possible in brothels on shore but also on board a ship. Due to the long voyages officers were permitted to take their wives aboard. So far it is not clear how often that was the case. In some instances also 'petty officers' like paymasters, coopers, sail makers and boatswains were allowed to take their wives on board the ship. From VOC ships it is reported that between 1652 and 1723 as much as

57 Anderson: 'The Ferringees are flying – the ship is ours!', pp. 144–145, 147.
58 Ibid., pp. 148–150.

40 children were given birth.[59] This was in stark contrast to the ordinary seamen. Though many of them were married, their women had basically no chance to join their husbands. For the privileged classes including passengers on board a ship sexual intercourse was possible whilst the majority of the crew had basically no opportunity of sexual activities.

Of course there were plenty of opportunities for same-sex activities. The situation on board a ship may not have differed from the situation in jails, cloisters and barracks. Homosexual activities on ships are seldom reported, yet they took place. To what extent, however, remains unknown due to the lack of sources and / or due to the lack of systematic scientific research. It seems that such activities were tolerated by the members of the crew as well as the officers. Sailors or officers were rarely accused of sodomy (the legal term for male anal intercourse or sexual intercourse between humans and animals), a capital crime in European countries.[60] Offenders on ships were kept in confinement until their trial ashore. Homosexual intercourse was not bounded to a specific deck but possible between decks and thus also classes.[61] Apart from the few reported cases of sodomy it seems very likely that the physically weaker young boys of the crew were sexually abused and in cases also raped by the elder sailors as well as the officers.

The situation of women was very serious on a ship, as cases of kuli passages from British India to the Caribbean demonstrate. A minority of women worked as indentured labourers on the plantations in Jamaica, Trinidad and Tobago as well as on Mauritius. To avoid the infamous regular rape of African slave-women on the Middle Passage by members of the crew, on kuli-ships special legislation attempted to prevent such sexual abuse. However, neither sexual segregation, spatial

59 Dok.. 19: Frauen an Bord: Prostituierte und Travestiten, Passagierinnen und Seemannsfrauen auf englischen und niederländischen Schiffen, In: Eberhard Schmitt (ed.): *Indienfahrer 2. Seeleute und Leben an Bord im ersten Kolonialzeitalter* (= Dokumente zur Geschichte der europäischen Expansion, vol. 7). Wiesbaden: Harrassowitz 2008, pp. 112–121, here pp. 114–115.

60 Dok. 22: Homosexualität an Bord. Das Kriegsgerichtsverfahren gegen John Morris und William Savage (1798), In: ibid., pp. 130–137; On homosexuality in the eighteenth century see Georgy E. Haggerty: *Men in Love. Masculinity and Sexuality in the Eighteenth Century*. New York / Chichester: Columbia University Press 1999.

61 Degryse: Social Conditions and Tensions on Board the Eighteenth Century East India Ships, pp. 343–344.

organisation, hierarchy of the officers and the crew nor the putative separation of the different ethnicities (African, Indian, European) on board the ships protected Indian women from being sexually molested and / or raped. Seen from a then contemporary European Orientalist perspective, 'coloured' women were *per se* regarded as sexually lascivious, which made it even more difficult for them to prove of having been raped. In addition, white men (captains, judges etc.) regarded lower class Indian women as particularly promiscuous and as prostitutes, and thus rape was hardly possible according to contemporary European legal constructions.

Indian women did not have a safe and sound journey to the colonies in eastern Africa and the Caribbean. Due to the separate quarters for men and women on board a kuli ship, sailors as well as officers had direct access to the women's deck. Kuli-women were preyed upon by males of all the above mentioned ethnicities, many of whom waylaid the women in the area of the toilets. Several cases of indentured women being raped during the voyage have been reported. Yet one case is of special interest, as it involved two members of the crew in their early twenties of African and European origin raping, as well as the doctor investigating the women who later died. In the end the case was dropped since the stereotype of promiscuous Oriental women and the belief that Indians (as every Oriental) were unreliable witnesses prevented any convincing evidence.[62] In some instances doctors abused their privileged position harassing, molesting and even raping indentured women from British India.[63]

Two observations with regard to these examples are of particular interest. First, it is evident that the limits of space and protection provided for a niche in which rape took place perverting sensual and sexual attraction into sheer aggression and subjugation of physically weak women and boys and ethnically different people. Second, despite ethnic separation aboard a ship with respect to working and living spaces and despite racial prejudices, common action against ethnically different women was possible. Cramped living conditions, lack

62 Verene A. Shepherd: *Maharani's Misery. Narratives of a Passage from India to the Caribbean.* Kingston: University of the West Indies Press 2002, pp. 33–79.

63 Marina Carter: *Voices from Indenture. Experiences of Indian Migrants in the British Empire.* London / New York: Leicester University Press 1996, pp. 94–96.

of individual space, hard working conditions and separated women quarters created the flammable mixture which exploded as soon as opportunity was produced. However, to protect a ship's hierarchical society under the authoritarian regimen of a captain and, more generally, to uphold the liberal-bourgeois European standard when it came to a case, the female victim was victimised thrice: when raped by a man or men, when examined by the ship's doctor and questioned by the captain and finally, when the sentence was rendered by the judge.

Fun and Fear

On board a ship there was not much time and space for fun. Work was almost perpetual, at least for the members of the crew. When men were not at work they either laid in their hammocks, singing or telling tales, or they played cards. Sometimes they were "sporting", catching fish which also enriched the generally dull diet.[64] In contrast to the few moments of fun there was plenty of fear. Many newly recruited kulis and lascars, before going on board a ship, had no clue what to expect. The many of rumours about the *kala pani* (lit.: black water) and about strange things going on at sea caused a lot of fear. On board a ship lascars were quite frightened, since most of them had been recruited from remote villages in eastern Bengal, in particular Sylhet District, or originated from the mountain regions of the Panjab. One lascar stated that:

> They took me down, inside the ship, such a strange place. I was frightened, because I had never seen anything like that before. Then the sea started to go [...] like that [...] *oh my God!* [...] I think for six or seven days I didn't have any cup of tea or food or anything. Still I had to work. I swore to God, when I get back I am never going to leave my home and family again. The other men looked after me, they said, 'Don't worry, after a few days the sea will go smooth.' It was true, I was alright when I got used to it.[65]

Crossing the ocean, *kala pani*, caused much fear among the indentured labourers as they had been told or had heard that they would lose their 'caste' (*jati* and *varna*). There has been a huge debate on the role of the *kala pani* for indentured labourers traversing the ocean. Until the

64 Anderson: 'The Ferringees are flying – the ship is ours!', p. 160; Sutton: *The East India Company's Maritime Service, 1746–1834*, p. 134. *Vide* also Hoffmann: *Ostindische Voyage 1671–1676*, pp. 13–14.

65 Adams: *Across Seven Seas and Thirteen Rivers*, p. 28.

middle of the nineteenth century, only lascars and merchants had been crossing the ocean. The issue became prominent during the second half of the nineteenth century when millions of labourers from the Indian Subcontinent were engaged as kulis to work on the plantations in the Caribbean or in Mauritius, in South, and in East Africa. The emerging Indian middle class as well as the Hindu reform organisations maintained that according to the Vedic scriptures it was forbidden to traverse the ocean. Yet in this case the Vedic text expresses more a general aversion than a strict interdiction as until the middle of the nineteenth century crossing the ocean was not explicitly prohibited. For example, with respect to merchants shipping their commodities over the ocean no argument was brought forward against the sea voyage.[66]

Crossing *kala pani* was made an issue in an emerging Indian public sphere when Indians belonging to the middle class and elite started travelling to England for university education. The *kala pani* issue became part and parcel of a Hindu re-assertion and, in fact, invention of a Hindu tradition commencing in the 1870s. To strengthen their argument, Hindu revivalists (or nationalists) referred to what they claimed to be the ancient Hindu scriptures.[67] It is obvious that the *kala pani* issue rather was a problem of the Hindu nationalists than of the merchants, students and indentured labourers. However, although many of the recruited kulis belonged to the *adivasi* and *dalits*, who are per definition not members of the brahmanic *varna* order, the public debate added to and disseminated anxiety and fear among the kulis who had never seen the ocean let alone traversing it in the bottom of a ship.[68]

Recruited kulis often did not know that their destination was overseas. Even when they realised that their future working place would be located overseas, making an embarkation on a vessel necessary, the length of the passage lasting several weeks on the Indian and Atlantic Ocean was a shock to many.[69] When arriving at the depots in

66 Susmita Arp: *Kālāpāni: Zum Streit über die Zulässigkeit von Seereisen im kolonialen Indien.* Stuttgart: Steiner 1998, pp. 5–46.

67 Ibid., pp. 46–85.

68 Marina Carter / Khal Thorabully: *Coolitude. An Anthology of the Indian Labour Diaspora.* London: Anthem Press 2002, p. 37.

69 Ibid., pp. 38–39.

Madras and Calcutta many of the recruited kulis believed that they were bound to some plantation on the subcontinent. Realising that he would be shipped over the ocean, Abheeram, a peasant from Orissa, recounted:

> […] on my way to the ship I refused to go but the Duffadars [recruiting agent] told me that there was no fear as the Mauritius was only eight days' journey. I told the Duffadar Bulram Singh that I would never venture going on shipboard. He told me to keep quiet and to tell the Sahebs if questioned that I was going on my own free will and accord and to convince me that there was no fear. He the Duffadar said that he would go on the same ship with me to the Mauritius.[70]

During the early years of indentured labour recruitment it was not uncommon that kulis 'deserted' from the depots or on the way from the depots to the ship. Men jumping overboard features regularly in the ships' logs. The risk of losing kulis was high when ships docked in a port en route to the Caribbean, most likely on a western Indian Ocean island or in a South African harbor. Some kulis went on hunger strike and in some instances rebellions had to be suppressed by force.[71] The reason for rebellion was not the status as a kuli, but the conditions aboard: lack of space, insufficient and bad food, inadequate hygiene and sanitation, none or deficient medical care. In addition, the habit of throwing dead people simply overboard caused horrors among the kulis with their funeral traditions and maybe heightened the willingness to protest and rebel.[72]

Food and Drink

Food aboard was fairly basic and from time to time also very inadequate—not only for the sailors, soldiers, slaves or convicts. Sometimes even the privileged passengers, on a long lasting voyage from Europe to India, complained about the food's bad quality.[73] In general, however, quality, opulence or frugality of the diet represented the social hierarchy of the society at home. The captain's table was often heavily

70 Carter / Thorabully: *Coolitude*, p. 37.

71 Ibid, p. 39.

72 Ibid., p. 41.

73 Christoph Adam Carl von Imhoff / Friedrich Wilhelm von Imhoff: Zwischen der Linie und der Insul Ceilon, 18ter August 1769. In: Gerhard Koch (ed.): *Imhoff Indienfahrer. Ein Reisebericht aus dem 18. Jahrhundert in Briefen und Bildern*. Göttingen: Wallstein 2001, pp. 104–105.

laden with meat, fish, vegetables and fruit for him and those having the privilege to share his hospitality. The situation below deck was quite different. Sailors, soldiers, slaves and convicts received only basic, not very nutritive food often consisting of dried meat—beef and pork—or dried fish with biscuits, cheese, beans, and beer. In the second half of the seventeenth century an order of a VOC ship stipulated that each member of the crew should receive half a pound bacon, one pound salted meat, seven pounds rice or four pounds of rusk (*zwieback*), half a "Maaß" of wine and one "Maaß" of brandy or arrack as well as nine "Maaß" of water.[74] Whether the rules and rations were obeyed by the captain remains unknown.[75]

Dried fish and rice was the diet of most of the lascars employed on the sailing ships as well as the steamers of the nineteenth and twentieth century. Additionally, the fresh water had often turned foul thus being hardly potable. Yet one has to keep in mind that except fresh water the vast majority of people on land did not have a daily diet including fish or meat, whether fresh or stinking.[76] Ships bound to East Africa or the East Indies stopped over in Bahia on the Brazilian coast, at St Helena, at the Cape of Good Hope, and Mauritius to replenish the diminishing food and water storages. Though the quantities of victuals suggest a sufficient supply with nourishing and filling food, it may be assumed that much of it rotted on board a ship due to the climate and storage below deck, as well as the lack of preservation possibilities. The vivid description of chasing rats on board a ship indicates the devastating effects rats and mice had on food storages and, at the same time, the

74 Hoffmann: *Ostindische Voyage 1671–1676*, p. 117. In Hessen, one "Maaß" was roughly two litres, cf. Johann Friedrich Hauschild / Georg K. Chelius: *Vergleichungstafeln der Gewichte verschiedener Länder und Städte: Nebst der neuesten Verordnungen uns Untersuchungen über Maße u. Gewichte, wie auch mehreren Beiträge zur Berichtigung der Gewichtskunde.* Frankfurt am Main: Verlag der Jägerschen 1836, p. 107.

75 Descriptions on the nutrition of the crew vary but basically food consisted of the same sort of meat, fish, rusk, beans, beer, rum and water rations, cf. Johann Gottlieb Worm: *Ost-Indian und Persianische Reisen, oder: Zehenjährige auf Groß Java, Bengala und in Gefolge Herrn Joann Josuae Kotelär, Holländischen Abgesandtens an den Soghi in Perisien, geleistete Kriegsdienste*, ed. by M. Crispinum Weise. Dreßden / Leipzig 1737, pp. 19–27.

76 Rediker: *Between the Devil and the Deep Blue Sea*, pp. 2–3; Davis: *The Rise of the English Shipping Industry*, p. 145.

story is telling about the sanitary conditions aboard a ship at the beginning of the eighteenth century.[77]

Apart from that it is also easy to imagine that in case of food shortages the crew, the soldiers as well as the slaves and convicts of the lower decks, received smaller rations than the privileged passengers and officers of the upper deck. Because of the unequal distribution of nutritious food the chance for wealthier passengers to survive the exhausting passage was by far greater than for the rest of the ship's population. In some instances it is reported that sailors deserted ships because the captain had shortened food rations to save money and enhance his margins. Sometimes captains had to reduce food rations in order to make the food supply last through a voyage that was prolonged due to contrary winds, storms or currents.[78] Once ashore, the physical condition of the passenger determined the chances of survival in the tropical climate to a large extent. Better nourished passengers were generally fitter to survive tropical diseases than weak and disease-ridden members of the crew, the military and the captive workforce.

The lascar crews' food was even more basic, mostly consisting of dried fish or dried meat and plain rice. Depending on the contract, lascars were obliged to buy the rice from their meager income suggesting that the rice was of low quality and therefore neither filling nor nutritive. Food was not only inferior but often inadequate. It seems that this simple diet did not change over the centuries as reports from the eighteenth and biographical notes from the twentieth century mention the same kind of food: daily rations of fish and rice. Some personal reports mention a better diet comprising dried fish, curry, *dal* (pulses) and rice.[79] Food may still have been very basic, yet pulses are more nutritive. However, the food rations of lascars employed on ships of the British merchant fleet were far cheaper than that for European

77 See-Stromer Jack. Die großmäulige Geschichte meines Lebens. Erzählt von Kapitän John Cremer in seinem achtundsechzigsten Jahr 1768. Hgg. von R. Reynell Bellamy. Deutsch von Erich Franzen. Berlin: S. Fischer 1936, pp. 65–66. Dok. 35: Der Unterkaufmann Jean François Michel über die Äquatorialtaufe auf dem Chinafahrer *Burg von Emden* (1752), In: Schmitt (ed.): *Indienfahrer 2*, pp. 238–240.

78 Rediker: *Between the Devil and the Deep Blue Sea*, pp. 101, 127.

79 Adams: *Across Seven Seas and Thirteen Rivers*, p. 30.

sailors.[80] More detailed information about the diet on a ship is available with regard to convict transportation. Basically convicts received the same food and the same rations as lascars. Apart from rice, fish and *dal*, small rations of tobacco, betel, ghee, tamarind, chilies, pepper, garlic and onions were part of the rations.[81]
According to personal statements, kulis received rice and *dal* twice a day. In another instance it has been stated that every kuli was given two bottles of potable water each day. Even in hot weather or when kulis were thirsty they did not receive additional water rations. The above mentioned Abheeram also stated that "[a]s long as I remained on board I lived on the Chorah [mashed peas] that I had got and occasionally on a small quantity of dry rice."[82] Eating cooked food was especially for high-*varna* Hindus an often insurmountable barrier. Many Hindus therefore refused to eat cooked food on board when cooks or companions were not of his *varna*. The voyage as well as food conditions on board traumatised many kulis of which some begged to be allowed to go home rather living as a beggar than being forced to eat the unclean food aboard the ship.[83]

Life and Limb

Sailors and Lascars

In the age of the sailing ship life and limb of a sailor and even more so of a lascar was cheap. Apart from the many diseases sailors contracted during their journey either ashore in African, South American and Asian ports or on board a ship, sailors and lascars had to contend with an extensive range of disabilities and afflictions which followed from their physical work. Frequently lifting and pulling, sailors and lascars were susceptible to hernia or the "bursted belly" as they called it. Easily a finger could be lost to a rolling cask, an arm or leg broken by shifting the cargo or fixing the cargo during heavy weather or a hand burned in tarring ropes. In particular during a storm the risk of a sailor to be swept overboard by a high wave, hit by falling ropes, beams and other heavy material, was fairly high. Some were crushed between the

80 Ahuja: Subaltern Networks under British Imperialism, p. 40.
81 Anderson: 'The Ferringees are flying – the ship is ours!', p. 154.
82 Adams: *Across Seven Seas and Thirteen Rivers*, p. 149.
83 Carter / Thorabully: *Coolitude*, p. 37.

mast and the side of the ship, some falling down the rigging, severely hurting themselves or, more likely, dying.[84]

In the 'Age of Steam' life and work of lascars took place below deck with dim light and in sticky air. In fine weather the lascars had their food on deck sitting in a circle on the floor and snatching some fresh air away from the engine-room fumes. Being away from the engine-rooms meant, being away from greasing and oiling the machinery, supplying the furnace with coal and disposing of the ashes or trimming the coal to keep the fire under control. Shifts lasted for four hours, followed by eight hours off.[85] Due to the hard and dense labour, accidents were frequent, illness and death omnipresent. This situation improved only during the nineteenth century. As mentioned above, the owners of British shipping companies relied more on lascars in the 'Age of Steam' since they regarded them more suitable for the working conditions in the engine-room than British sailors. Lascars were supposed to better endure the extremely high temperatures close to the furnaces and could therefore withstand the heat under a steel deck on the Arabian Sea.[86]

The lascar's life and work was organised along racial categories. To avoid conflicts between the shipmaster and the chief engineer, the crew was strictly divided into a "deck" and "engine" department. The "purser's" or "steward's" department made up the third compartment. Lascars were recruited from different regions and different ethnic backgrounds in South Asia and in harbours around the Indian Ocean. Crew members with different duties and working areas would therefore hardly be able to communicate among each other. Thus mixed crews combining lascars from Africa, Arabia and Asia were highly undesirable. However, cultural homogeneity among each of the three lascar compartments was also desired. The limited homogeneity as well as the strict segregation provided for a better control of the lascar workforce and limited the interaction or solidarity between them, which was so significant for the seamen during the 'Age of Sail'.[87]

84 Rediker: *Between the Devil and the Deep Blue Sea*, pp. 1–2, 92–93.

85 Adams: *Across Seven Seas and Thirteen Rivers*, p. 30.

86 Ibid., p. 22.

87 Ahuja: Subaltern Networks under British Imperialism, p. 47; Rediker: *Between the Devil and the Deep Blue Sea*, *passim*.

Convicts

As mentioned above, the space allotted to the convicts was about the same to lascars and Indian soldiers. However, convicts from South and Southeast Asia were accommodated between decks, with their provisions in the hold and amidst the cargo of the ship, amongst many other commodities such as cotton bales, reels of silk, chests of opium, rattan canes and sacks of betel nut. To reduce costs, ship owners often refused to take a doctor aboard a ship. Sometimes the lack of medical care and the lack of space turned the voyage into a horror-trip as in the case of the *Atlantic* transporting convicts from India to Singapore in 1857. Of the 131 prisoners nineteen died on board whilst another forty were still in hospital two months after disembarkation. Convicts stated that they had been crammed into a poorly ventilated room below deck, space being limited to six by half a foot which made it impossible to sit and to lie properly. Men suffered from fever, diarrhea and dysentery. In addition, convicts had been allowed to wash only once during the journey. Their food rations had been totally insufficient and they suffered from water shortages.[88]

Other instances are reported where convicts were allowed to wash their clothes and bedding only once during a voyage. In most instances convicts were forced to sit, sleep, eat and answer the call of nature on the very same spot. In addition to these 'uncomfortable' living conditions, fetters and irons caused severe problems. Generally convicts were put in chains whilst below deck. In some cases they were also chained from waist to wrist and neck. Often fetters caused sores and wounds on the convict's ankles, legs and wrists. When left untreated, which was very likely without a doctor aboard, injuries could lead to gangrene and, worse, even death. At the same time shackles and fetters were not in a good condition. In the humid and salty climate below deck, the iron corroded within a short time which enabled the convicts to free themselves and rebel turning the means of strict confinement into formidable weapons.[89]

Some of the convicts could obtain privileged positions. Despite the regulations prohibiting the work of convicts on board a ship some of them were integrated into the life on deck. The captain selected

88 Anderson: 'The Ferringees are flying – the ship is ours!', pp. 149, 157.
89 Ibid., pp. 157–160.

convicts according to their regional origin, social backgrounds, religion, skills, etc. Some were engaged as sweepers, cooks and even overseers over their fellow convicts. They may have also been engaged for loading and unloading cargo. During their duties, most of these convicts did not have to wear fetters. This brought them physically close to the members of the crew. In case of a rebellion started by the convicts alliances between crew and convicts were likely especially when some of the convicts were sailors by profession. Yet there are only few instances reported when sailors mutinied and prisoners rebelled together or sailors assisted convicts. At the same time, one has to bear in mind that the living conditions of sailors and prisoners aboard were quite similar which might have been another reason why the former supported the latter in cases of upheaval.[90]

Disease and Death

Soldiers

A report on the voyage of HM 73rd and 76th regiments in 1790 states that cleanliness was essential if the soldiers and recruits, in their cramped conditions, were to remain healthy. According to regulations, every week the officers examined all the soldiers' clothing and issued them with soap.[91] In the negative, this coincides with the critical observation of a French doctor aboard a VOC vessel bound to the East Indies stating that the soldiers aboard the ship were not particularly interested in their personal cleanliness.[92] On some of the troop-transporters horrible sanitary and medical conditions did exist as the case of the Hessian troops shipped to Madras indicates. After recruitment the soldiers of the two regiments were taken to Hameln (Hamlin) where they were drilled and disciplined, formed into the 15th and 16th regiments and sent to Stade where the troops were embarked for England. Due to stormy weather the hatches had to be closed. Insufficient ventilation and dampness soon caused the outbreak of typhoid, lead-

90 Anderson: 'The Ferringees are flying', pp. 151–152, 162.

91 Sutton: *The East India Company's Maritime Service, 1746–1834*, p. 173.

92 Louis R. Rouppe: De Morbis Navigantium. Lugduni Batavorum [Leiden] 1764, Dok. 44: Mangelnde Hygiene, ungesunde Ernährung, harte Arbeit: Zur gesundheitlichen Gefährdung auf See, In: E. Schmitt (ed.): *Indienfahrer 2,* pp. 300–313; part c, pp. 309–313.

ing to almost one hundred men dying during the passage to England or during the following weeks ashore.[93]

Only three out of four ships reached Plymouth. As health conditions of the soldiers were lamentable, half of them had to be taken to hospital. The lost ship returned to Ritzebüttel, a small harbour on the Elbe close to Cuxhaven, where it hibernated, frozen and with soldiers on board the ship. Lack of adequate clothing and blankets as well as lack of food caused dozens of deaths. Finally, the soldiers mutinied and left the ship to seek shelter in town. Only later all soldiers were taken to Stade where they stayed until March then sailing for Portsmouth. During the passage to India the troops had to suffer again. Stormy weather sunk one of the ships close to the Comoro islands where many seamen and soldiers died. When the ships reached Madras in April 1783, the magazine of one of the ships exploded causing the death of one officer and five soldiers. To replenish the loss, another two companies consisting of 400 soldiers were sent to Madras in 1785 and another 200 soldiers two years later. So far nothing is known about their sufferings and destinies.[94]

Mortality rates on ships were sometimes extremely high. This is in particular true for the crew, the soldiers and the slaves. On EIC ships on their way to the Indic mortality was 10 per cent in the seventeenth century, during the eighteenth century death rate sometimes could exceed 20 per cent. With respect to soldiers, mortality rates seem to have been rather low. In the second half of the eighteenth century it varied between 1.5 and 3 per cent. However, some instances indicate horrendous mortality numbers, when out of fifty-three officers and men on the *Osterly* thirty-three died onboard. In the same year out of sixty-one soldiers on board the *Worcester* forty-three died at sea.[95] The mortality on VOC ships was between 7 and 15 per cent between 1620 and 1775. In comparison to the mortality rate on shore, mortality on board a ship was twice the size.[96] On Portuguese ships bound to India mortal-

93 Diebitsch: Die kurhannoverschen Truppen in Ostindien 1782–1792, p. 90.

94 Ibid., pp. 90–91, 114.

95 Gilbert: Recruitment and Reform in the East India Company Army, 1760–1800, p. 92, fn. 7.

96 J. R. Bruijn: De personeelsbehoefte van de VOC overzee en aan boord, bezien in Aziatisch en Nederlands perspectif. In: *Bijdragen en Mededelingen betreffende de Geschiedenis der Nederlanden BMGN* 91 (1976), pp. 218–248, esp. p. 223. Between 1707 and 1775, in

ity among the crew was between 12 to 15 per cent. In a few instances it could amount to 35 per cent of all passengers on board a ship, most likely when the transport included a large number of slaves.[97] To a very large extent, the size of a ship and the amount of people being on board determined the death rate among its 'population'.

Lascars

From the end of the eighteenth century lascars were also employed on England-bound EIC ships. Under-nourishment, malnutrition and insufficient clothing increased the amount of illness amongst the lascars. Catching a cold was a rather negligible infection, pneumonia a severer one. From the end of the eighteenth century to the middle of the twentieth century, British officials opined that lascars were unsuited for service in cold climate—as European sailors were unsuited for hard manual work on ships. However, lascars were never given adequate clothing on their voyages to Europe, which is why the stereotype of the weak lascar physically unfit for work in cold climate became well established in the course of one and a half centuries.[98]

Due to the appalling heat close to the furnace in the engine-room many a lascar died from heat strokes, dehydration, or exhaustion. Others, being utterly desperate, threw themselves overboard. In many instances, lascars ran away from the ship when reaching a harbour. Yet, sometimes the captain employed watchmen to prevent lascars from escaping the ship only with little success. As one lascar remembered being in the engine-room in the middle of the 1930s,

> I was called a bunkerman, in the coal bunker [...] it was hot, oh yes, it was hot. We put coal in the boiler, and then it gets heat, and then the ship runs. It is a most difficult job, very hard and very hot too, many people died that heat. In my sea life, I knew hundreds of people who died.[99]

Bengal the mortality rate of Europeans was 58 percent (378 out of 645), most of them in their 20s, cf. Peter Marshall: *East Indian Fortunes. The British in Bengal in the Eighteenth Century*. Oxford: Oxford University Press 1976, table III, p. 218. Among the civil servants of the EIC mortality rate reached up to 20 percent, Herbert Hilder: The East India Company's Regulation of its Servants – A Study in Failure: Bengal in the 1760s. In: *Bengal Past and Present* 97 (1978), pp. 1–20, esp. p. 19, fn. 13.

97 Lopes: Seamen, Soldiers and Crew of the *Carreiro da India* in the Second Half of the Eighteenth Century, p. 137.

98 Home: *Merchant Seamen, passim.*

99 Adams: *Across Seven Seas and Thirteen Rivers*, pp. 22, 149 (quote), 130.

Lascars also died as a consequence of accidents, especially falls, burns from steam and gas explosions, avalanches of coal as well as dysentery. In some cases the combination of illness and injury proved fatal.[100] The lot of the kulis was by far not better. Despite sanitary regulations promulgated by the British Government of India during the 1860s and 1870s, their mortality rate could easily exceed 4 per cent during the second half of the nineteenth century. To counterbalance human losses to be expected during the sea voyage, captains and contractors overloaded the ships up to 10 per cent of the stipulated maximum number of passengers. That, however, was another reason for the outbreak of diseases like cholera, typhus, dysentery etc., and numbers of sick and dead people rising accordingly.[101]

Seamen

During the seventeenth and eighteenth centuries almost half of the sailors recruited for service on deep-water ships died at sea. The West African coast was deadly for European and American crews who visited its shores to pick up slaves. The sailors contracted malaria, yellow fever, scurvy and dysentery. It is generally agreed upon that Anglo-American seamen died in roughly the same horrendous proportions, and occasionally in even greater ones, as did the slaves themselves. In 1714, for example, the surviving crew members of the *Florida* tossed the dead bodies of their comrades over board at night to disguise the rapidly dwindling numbers of seamen from the slaves whilst slave corpses were thrown overboard during the day.[102]

As the voyage to the East Indies was much longer than a trans-Atlantic passage, many seamen had already been infected with dysentery or typhus when half way down the African coast. Reaching the Indian Ocean, many sailors but also soldiers showed traces of scurvy—or even more than that. In East African and Indian port cities seamen tried to recover from the diseases, sometimes taking lots of alcohol as a medical treatment which of course worsened the situation—and sometimes acquiring new diseases. On the way back, St Helena seems to have been haven and heaven for the sick and worn out sailors. Their

100 Ewald: Crossers of the Sea, p. 88.

101 Hugh Tinker: *A New System of Slavery. The Export of Indian Labour Overseas 1830–1920*. Oxford: Oxford University Press 1974, pp. 116–119.

102 Rediker: *Between the Devil and the Deep Blue Sea*, pp. 32–33, 46–48, cf. fn. 89.

diaries and journals mention the island some 1,000 miles off the shores of southern Africa as a place for quick recovery from diseases. Francis Rogers, for example, noted that many men "just [nearly] dead with the scurvy and other diseases [...] were carried ashore here, recovered to a miracle, rarely any dying though never so ill when brought ashore."[103] In the Indian Ocean, death and disease was as omnipresent as on the trans-Atlantic passages, however, whether mortality rates were larger here than there remains to be proven.

Convicts

Mortality rates of convicts on board British ships seem to have been rather low. On board 215 convict-ships sailing from Bengal between 1793 and 1848 only 25 prisoners died. During the 1840s and 50s, on some convict ships mortality rates were extremely high, going up to 12 per cent. Official reports stated that the high number of dead convicts was due to the journey itself. It was maintained that on embarkation all convicts had been in healthy condition. However, in some instances it seems that despite serious illness convicts refused to eat or take medicine. Other reports show that convicts with venereal diseases entered ships without a doctor on board. Many of the convicts suffered from debilitating seasickness adding to their misery and desperation. In 1827 seven convicts refused to board the *Constance* because they deemed the lower decks hot enough to kill them.[104]

This rather low number of mortality on British ships is contrasted by Portuguese transportation. On Portuguese ships, prisoners comprised most of the people travelling to India. In many cases more than 200 prisoners or 80 per cent of the people on board were convicts being transported to Goa. Among the prisoners mortality was extremely high. In the second half of the eighteenth century at least 1730 prisoners died en route to India, which represents roughly 45 per cent of all Portuguese prisoners carried to India.[105] In comparison, for example, mortality among the white settlers migrating to North America in the

103 Rediker: *Between the Devil and the Deep Blue Sea*, pp. 39–41.

104 Anderson: 'The Ferringees are flying–the ship is ours!', pp. 153–157.

105 Lopes: Seamen, Soldiers and Crew of the *Carreiro da India* in the Second Half of the Eighteenth Century, pp. 136–137. Between 1750 and 1800 almost 14,000 persons travelled to the Portuguese possessions in South Asia, ibid., p. 136. However, it seems not clear whether that included African slaves.

middle of the nineteenth century figured between 1 and 2 per cent whilst that of the European convicts shipped to Australia amounted to roughly 11 per cent during the last decade of the eighteenth century. Numbers dropped after faster ships reduced the voyage between Europe and Australia from 190 to 169 days falling to less than 1 per cent from the 1820s onwards.[106]

Slaves

Lots of calculation took place during the last fifty years to get to know the actual number of slaves shipped from Africa to the Americas and the Caribbean. With respect to the Indian Ocean such meticulously elaborated estimates do not exist. Some publications try to compare the Atlantic mortality rate among the slaves between 1500 and 1900 with the mortality rate of "Muslim" slave trade in the Indian Ocean between 1200 and 1900 claiming that mortality in the latter trade was at least as high as in the former.[107] Yet comparisons along 'religious boundaries' are unsatisfying as they repeat the age-old argument of a despotic Islamic Orient including slavery and slave trade thus, again, contributing to a well established Orientalist discourse. Instead one should have a look at the number of slaves aboard ships in the Indian Ocean. Some examples may give an impression of the slave trade in the Indian Ocean during the eighteenth and nineteenth centuries.

To increase their profit margins many European captains and ship owners put as many slaves as possible on board a ship, sometimes more than 300. As it was the case with the soldiers and the convicts the density of people on board a ship as well as the length of the voyage determined the sanitary conditions, the outbreak of diseases and, finally, the mortality rate. Sometimes the number of dead people could exceed 25 per cent.[108] Between 1773 and 1810 the voyage from Mauritius to Zanzibar or Mozambique caused the death of sometimes a quarter of the slaves. During the by far shorter passage between the eastern coast of Madagascar to Mozambique or Zanzibar 'only' 10 per cent of the slaves died.[109] At the end of the eighteenth century slave

106 Klein: *The Atlantic Slave Trade*, pp. 134–136.

107 Summarised in Mann: *Sahibs, Sklaven und Soldaten*, p. 124.

108 Marina Carter: Indian Slaves in Mauritius (1729–1834). In: *Indian Historical Review* 15 (1987), pp. 233–247, cf. p. 237.

109 Deryck Scarr: *Slaving and Slavery in the Indian Ocean.* Houndmills: Macmillan 1998, p. 34.

transports on ships in the Indian Ocean were quite similar to that of the Atlantic Ocean when, after the outbreak of epidemic diseases, the mortality rate could sometimes exceed 20 per cent.

The extremely high mortality rate on Portuguese slave-ships—in some cases up to one-third during the seventeenth century—must also be seen against the background of the outbreak of diseases like plague or dysentery.[110] During the eighteenth century mortality on the slave ships from eastern Africa to Portuguese Daman and Diu on the western coast of India was 18 per cent, which also means double the average percentage of the Atlantic Middle Passage.[111] The only mortality rate, which exceeded that number was the slave trade between Mozambique and Brazil. Due to the extreme long voyage around southern Africa and the cold weather of the Cape region and the southern Atlantic, in a few cases mortality went up to 28 per cent.[112] *Dhaus* carried between 80 and 100 slaves over the Mozambique Channel, in some instances also up to 250 slaves during the heyday of slavery and slave trade in the middle decades of the nineteenth century.[113]

The conditions on board these ships must have been the same as on European ships crossing the Atlantic. To keep the mortality rate low Arab and African slavers tried to keep the journey short. This might also have been the reason why they established the Seychelles and the Comoros as secret reloading sites trying to hide their trade from British ships patrolling the Eastern African coast.[114] So far nothing is known about the sanitary, health, food and clothing conditions on such ships. In case of a large number of slaves on board a ship food rations and sanitary conditions might not have been the best and, in some contrast, might have been better on smaller boats. Further research is needed to shed some more light on this neglected field.

110 Jeanette Pinto: *Slavery in Portuguese India: 1510–1842.* Bombay / Delhi / Nagpur: Himalaya 1992, pp. 35, 40–44.

111 Pedro Machado: A Forgotten Corner of the Indian Ocean: Gujarati Merchants, Portuguese India and the Mozambique Slave-trade, c. 1730–1830. In: Gwyn Campbell (ed): *The Structure of Slavery in Indian Ocean Africa and Asia.* London: Cass 2004, p. 20.

112 Edward A. Alpers: *Ivory and Slaves in East Central Africa.* London: Heinemann 1975, p. 211.

113 Gwyn Campbell: Madagascar and the Slave Trade, 1810–1895. In: *Journal of African History* 22 (1981), pp. 203–227.

114 Raymond W. Beachey: *The Slave Trade of Eastern Africa.* London: Collings 1976, pp. 68–70.

Conclusion

First of all, living conditions on board the ships in the Indian Ocean when carrying slaves, kulis or soldiers in large numbers seem to have been almost the same in the eighteenth century. Often food was insufficient due to the long journey, health care negligible if existing at all, sanitary conditions horrible in comparison to present-day or even contemporary standards. What is striking is the fact that the mortality rate among the slaves and the seamen in the Indian Ocean was as high as that of the Atlantic. Although the number of passengers like slaves and soldiers was definitely lower than on Atlantic ships, the major reason for the similar number of dead persons was the rather long voyage in the Indian Ocean. Mortality among the privileged classes on board European ships sailing to the Indian Ocean was much less though exact numbers are still missing. Yet this would at least indicate that class distinction and racial differentiation was the reason for high mortality numbers among the lower deck passengers.

Secondly, in contrast to the Atlantic maritime world, the maritime culture of the vessels in the Indian Ocean was different. In many cases the boundaries among the members of an Asian ship's crew seem to have blurred the categories of slave, freedman, seaman, lascar and wage worker in the nineteenth century thus forming some kind of horizontally structured class of maritime labourers of the lower decks. At the same time, racially defined hierarchy between the European members on the one hand and the Asian and African members of a European ship's crew, on the other hand, definitely increased, Asian lascars and European sailor again becoming different legal categories like in the sixteenth and seventeenth centuries. In any case the gap between the lower deck crew and upper deck passengers widened. Whether the maritime culture shared many features with the plebeian culture like language, beliefs, non-believing, social backgrounds etc. as was the case with the North Atlantic maritime culture remains to be proven.[115]

Thirdly, the maritime culture of the subalterns was dominated by an omnipresent labour regime, which determined the life aboard ship. Except the slaves and the convicts and in some cases also the soldiers, every member of the crew was bound to work all around the clock in shifts regulated by the strike of the bell. Slaves and most of

115 Rediker: *Between the Devil and the Deep Blue Sea*, pp. 201–202.

the convicts remained chained throughout the whole passage. Hardly anything is known about the situation of soldiers and in particular when whole regiments were to serve in Eastern Africa, South Asia or Southeast Asia. This gap of knowledge still needs to be filled. Likewise, more research is needed to investigate into the sexual life and even love on board of ships. Piracy, in many cases the result of a crew who mutinied, seems to have provided a niche in which same-sex relationships were tolerated to some extent.[116]

Fourthly, "no peace beyond the line" implicates that treaties signed between European powers were not valid 'beyond the line', which opened and defined the maritime space for buccaneering, freebooting and piracy. It also indicated that the old debate about *mare clausum* and *mare liberum* was still on the agenda of European maritime powers. Likewise it pointed to some kind of permanent war status existing in the Caribbean, South America, Africa and Asia yet not in the Northern Atlantic and Europe where treaties of amity bound sovereigns and governments. And it implicated as well that civilisation and culture differed. Rules, laws and values respected in Europe were often suspended 'beyond the line'. Subaltern life on ships seems to mirror this gap between a culturally constructed civilised and an uncivilised world even when there was no war. Contracts of seamen, i. e. lascars stipulated harsher working conditions than those of European sailors, indentured labourers' rights were more limited than contemporary English industrial workers' rights and African as well as South Asian women were basically regarded as sexual objects and fair game left unprotected by existing laws and regulations. To conclude, if life on board the ships of the Atlantic was "between the devil and the deep blue sea", in many instances life on board ships in the Indic seems to have been hell on the water.

116 Burg: *Sodomy and the Pirate Tradition*, pp. 107–138.

Bibliography

Adams, Caroline: *Across Seven Seas and Thirteen Rivers. Life Stories of Pioneer Sylheti Settlers in Britain.* London: THAP Books 1987.

Agius, Dionisius A.: *In the Wake of the Dhow.* Reading: Ithaca 2002.

Ahuja, Ravi: Subaltern Networks under British Imperialism. Exploring the Case of South Asian Maritime Labour (c. 1890–1947). In: Jan-Georg Deutsch / Brigitte Reinwald (eds): *Space on the Move. Transformations of the Indian Ocean Seaspace in the Nineteenth and Twentieth Century.* Berlin: Schwarz 2002, pp. 39–60.

—: Lateinsegel und Dampfturbinen. Der Schiffsverkehr des Indischen Ozeans im Zeitalter des Imperialismus. In: Dietmar Rothermund / Susanne Weigelin-Schwiedrzig (eds): *Der Indische Ozean. Das afro-asiatische Mittelmeer als Kultur- und Wirtschaftsraum.* Wien: Promedia 2004, pp. 207–226.

—: Die 'Lenksamkeit' des 'Lascars'. Regulierungsszenarien eines transterritorialen Arbeitsmarktes in der ersten Hälfte des 20. Jahrhunderts. In: *Geschichte und Gesellschaft* 31,3 (2005), pp. 323–353.

—: Mobility and Containment. The Voyages of South Asian Seamen, c. 1900–1960. In: *International Review of Social History* 51 (2006), pp. 111–141.

—: Networks of Subordination – Networks of the Subordinated. The Ordered Space of South Asian Maritime Labour in an Age of Imperialism (c. 1890–1947). In: Ashwaini Tambe / Herald Fischer-Tiné (eds): *The Limits of British Colonial Control in South Asia. Spaces of Disorder in the Indian Ocean Region.* London / New York: Routledge 2009, pp. 13–48.

Alpers, Edward A.: *Ivory and Slaves in East Central Africa.* London: Heinemann 1975.

Anderson, Clare: 'The Ferringees are flying – the ship is ours!'. The Convict Middle Passage in Colonial South and Southeast Asia, 1790–1860. In: *Indian Economic and Social History Review* 42,2 (2005), pp. 143–186.

Arp, Susmita: *Kālāpānī. Zum Streit über die Zulässigkeit von Seereisen im kolonialen Indien.* Stuttgart: Steiner 1998.

Atwood, Rodney: *The Hessians. Mercenaries from Hessen-Kassel in the American Revolution.* Cambridge: Cambridge University Press 1980.

Balachandran, G.: Recruitment and Control of Indian Seamen. Calcutta, 1880–1935. In: *International Journal of Maritime History* 9,1 (1997), pp. 1–18.

Beachey, Raymond W.: *The Slave Trade of Eastern Africa.* London: Collings 1976.

Broeze, F. J. A.: The Muscles of Empire – Indian Seamen and the Raj, 1919–1939. In: *Indian Economic and Social History Review* 18,1 (1981), pp. 43–67.

Bruijn, J. R.: De personeelsbehoefte van de VOC overzee en aan boord, bezien in Aziatisch en Nederlands perspectif. In: *Bijdragen en Mededelingen betreffende de Geschiedenis der Nederlanden BMGN* 91 (1976), pp. 218–248.

Burg, B. R.: *Sodomy and the Pirate Tradition. English Sea Rovers in the Seventeenth-Century Caribbean.* New York: New York University Press 1995.

Campbell, Gwyn: Madagascar and the Slave Trade, 1810–1895. In: *Journal of African History* 22 (1981), pp. 203–227.

Carter, Marina: Indian Slaves in Mauritius (1729–1834). In: *Indian Historical Review* 15 (1987), pp. 233–247.

—: *Voices from Indenture. Experiences of Indian Migrants in the British Empire.* London / New York: Leicester University Press 1996.

Carter, Marina / Khal Thorabully: *Coolitude. An Anthology of the Indian Labour Diaspora.* London: Anthem 2002.

Davis, Ralph: *The Rise of the English Shipping Industry during the Seventeenth and Eighteenth Centuries.* Townbridge, London: Abbot 1972.

Degryse, Karel: Social Conditions and Tensions on Board the Eighteenth Century East India Ships. In: K. S. Mathew (ed.): *Mariners, Merchants and Oceans. Studies in Maritime History.* Delhi: Manohar Publishers 1995, pp. 341–346.

Dharampal-Frick, Gita: *Indien im Spiegel deutscher Quellen (1500–1750). Studien zu einer interkulturellen Konstellation.* Tübingen: Niemeyer 1994.

Diebitsch, Viktor von: Die kurhannoverschen Truppen in Ostindien 1782–1792. In: *Hannoversche Geschichtsblätter* 1 (1898), pp. 67–128.

Dixon, C. H.: Lascars. The Forgotten Seamen. In: Rosemary Ommer / Gerald Panting (eds): *Working Men who Got Wet.* St. Johns: Maritime History Group, Memorial University of Newfoundland 1980, pp. 263–281.

Ewald, Janet J.: Crossers of the Sea. Slaves, Freedmen, and other Migrants in the Northwestern Indian Ocean, c. 1750–1914. In: *American Historical Review* 105,1 (2000), pp. 69–91.

Foucault, Michel: Different Spaces. In: Id. / James D. Faubion: *Aesthetics, Method and Epistemology.* New York: The New Press 1998, pp. 175–185.

Gilbert, Arthur N.: Recruitment and Reform in the East India Company Army, 1760–1800. In: *Journal of British Studies* 15 (1975), pp. 89–111.

Graaff, Nicolaas de: Oost-Indise Spiegel. In: J. C. M. Warnsinck (ed.): *Reisen van Nicolaus de Graaff. Gedaan naar alle gewesten des Werelds beginnende 1639 tot 1687 incluis.* 's-Gravenhage: Nijhoff 1930 [1703].

Haggerty, Georgy E.: *Men in Love. Masculinity and Sexuality in the Eighteenth Century.* New York / Chichester: Columbia University Press 1999.

Hauschild, Johann Friedrich / Georg K. Chelius: *Vergleichungstafeln der Gewichte verschiedener Länder und Städte: Nebst der neuesten Verordnungen uns Untersuchungen über Maße u. Gewichte, wie auch mehreren Beiträge zur Berichtigung der Gewichtskunde.* Frankfurt am Main: Verlag der Jägerschen 1836.

Hilder, Herbert: The East India Company's Regulation of its Servants – A Study in Failure: Bengal in the 1760s. In: *Bengal Past and Present* 97 (1978), pp. 1–20.

Hoffmann, Johann Christian: *Ostindische Voyage 1671–1676*, reprint ed. by Günter E.Th. Bezzenberger. Kassel: Thiele & Schwarz 1983 [1680].

Home, William E.: *Merchant Seamen. Their Diseases and their Welfare Needs*. London: Murray 1922.

Huck, Stephan: *Soldaten gegen Nordamerika. Lebenswelten Braunschweiger Subsidiärtruppen im Amerikanischen Unabhängigkeitskrieg*. München: Oldenbourg 2011.

Hyslop, Jonathan: Steamship Empire. Asian, African and British Sailors in the Merchant Marine, c. 1880–1945. In: *Journal of Asian and African Studies* 44,1 (2009), pp. 49–67.

Klein, Herbert S.: *The Atlantic Slave Trade*. 8th ed. Cambridge: Cambridge University Press 2005 [1999].

Koch, Gerhard (ed.): *Imhoff Indienfahrer. Ein Reisebericht aus dem 18. Jahrhundert in Briefen und Bildern*. Göttingen: Wallstein 2001.

Lopes, Maria de Jesus Teodoro Dos Martires: Seamen, Soldiers and Crew of the Carreiro da India in the Second Half of the Eighteenth Century. In: K. S. Mathew / Satish Chandra (eds): *Ship-building and Navigation in the Indian Ocean Region, AD 1400–1800*. New Delhi: Munshiram Manoharlal Publishers 1997, pp. 132–139.

Machado, Pedro: A Forgotten Corner of the Indian Ocean. Gujarati Merchants, Portuguese India and the Mozambique Slave-trade, c. 1730–1830. In: Gwyn Campbell (ed.): *The Structure of Slavery in Indian Ocean Africa and Asia*. London: Cass 2004.

Mann, Michael: *Bengalen im Umbruch. Die Herausbildung des britischen Kolonialstaates, 1754–1793*. Stuttgart: Steiner 2000.

—: Der ungeliebte Krieg. Compagnie des Indes und East India Company als Kombattanten in einem globalen Konflikt, 1742–1763. In: Sven Externbrink (ed.): *Der Siebenjährige Krieg als (europäischer) Weltkrieg*. Berlin: Akademie Verlag 2010, pp. 99–126.

—: Migration – Re-migration – Circulation. South Asian Kulis in the Indian Ocean and Beyond, 1840–1940. In: Donna R. Gabaccia / Dirk Hoerder (eds): *Connecting Seas and Connected Ocean Rims. Indian, Atlantic, and Pacific Oceans and China Seas Migrations from the 1830s to the 1930s*. Leyden / Boston: Brill 2011, pp. 108–133.

—: *Sahibs, Sklaven und Soldaten. Geschichte des Menschenhandels rund um den Indischen Ozean*. Darmstadt: WBG 2012.

—: Circulation and Migration. In: Philippa Levine / John Mariott (eds): *The Ashgate Research Companion to Modern Imperial Histories*. Farnam / Burlington: Ashgate 2012, pp. 377–393.

Marshall, Peter: *East Indian Fortunes. The British in Bengal in the Eighteenth Century*. Oxford: Oxford University Press 1976.

Maurer, Hans-Martin: Das Württembergische Kapregiment. Söldner im Dienst früher Kolonialpolitik (1787–1808). In: *Zeitschrift für Württembergische Landesgeschichte* 47 (1988), pp. 291–307.

Pinto, Jeanette: *Slavery in Portuguese India. 1510–1842.* Bombay: Himalaya 1992.

Qaisar, A. Jan: From Port to Port. Life on Indian Ships in the Sixteenth and Seventeenth Centuries. In: Ashin Das Gupta / M.N. Pearson (eds): *India and the Indian Ocean 1500–1800.* Calcutta: Oxford University Press 1987, pp. 331–349.

Rediker, Marcus: *Between the Devil and the Deep Blue Sea. Merchant Seamen, Pirates and the Anglo-American Maritime World, 1700–1750.* 12th ed. Cambridge: Cambridge University Press 2007 [1987].

Scarr, Deryck: *Slaving and Slavery in the Indian Ocean.* Basingstoke: Macmillan 1998.

Schmitt, Eberhard (ed.): *Indienfahrer 2. Seeleute und Leben an Bord im ersten Kolonialzeitalter.* (= Dokumente zur Geschichte der europäischen Expansion, vol. 7). Wiesbaden: Harrassowitz 2008.

Shepherd, Verene A.: *Maharani's Misery. Narratives of a Passage from India to the Caribbean.* Kingston: University of the West Indies Press 2002.

Sutton, Jean: *The East India Company's Maritime Service, 1746–1834. Masters of the Sea.* Woodbridge: The Boydell Press 2010.

Tinker, Hugh: *A New System of Slavery. The Export of Indian Labour Overseas 1830–1920.* Oxford: Oxford University Press 1974.

Wenzelhuemer, Roland / Michael Offermann: Ships Newspapers and Passenger Life Aboard Transoceanic Steamships in the Late Nineteenth Century. In: *Transcultural Studies* 1 (2012), pp. 77–121.

Wiebeck, Erno / Hermann Winkler: *Segler im Monsun. Die Dau am Indischen Ozean.* Rostock: Neuer Hochschul-Schriften Verlag 2000.

Worm, Johann Gottlieb: *Ost-Indian und Persianische Reisen*, ed. by M. Crispinum Weise. Dreßden / Leipzig 1737.

Beyond Piracy

Maritime Violence and Colonial Encounters in Indian History

Derek L. Elliott / Sebastian R. Prange*

1. "What gives you the right?"

Presented with a captured pirate, Alexander the Great demanded to know what gave him the right to infest the sea with his robberies, to which his prisoner brazenly retorted: "The same as you do when you infest the whole world; but because I do it with a little ship I am called a robber, and because you do it with a great fleet, you are called an emperor."[1] This timeworn allegory, here in its famous rendering by St Augustine, has been invoked by political theorists from Cicero to Chomsky to draw attention to the knotty question of political legitimacy.[2] Taking the figure of the pirate to epitomize this issue is indeed most appropriate because defining piracy, like legitimacy, is more often than not a matter of perspective. Whereas older histories regarded piracy as an unproblematic category—Philip Gosse, in his pioneering *History of Piracy* of 1932, deemed the question of who was or was not

* Derek Elliott would like to acknowledge the generous support of a Shastri Indo-Canadian Institute research fellowship that made possible a visit to the Maharashtra State Archives. Both authors wish to thank the editors for their encouragement and support.

1 Cf. Augustine, Bishop of Hippo: *The City of God Against the Pagans*, ed. by R.W. Dyson. Cambridge: Cambridge University Press 1998, p. 148.

2 Marcus Tullius Cicero: *Cicero's Tusculan Disputations. Also, Treatises on the Nature of the Gods, and on the Commonwealth*, trans. from the Latin by C. D. Yonge. New York: Harper 1877, p. 433; Noam Chomsky: *Pirates and Emperors, Old and New. International Terrorism in the Real World.* Rev. ed. Cambridge, MA: South End Press 2002, p. vii.

a pirate a point that could be "easily settled"—more recent scholarship has come to acknowledge that "who is a pirate is often very much in the eye of the beholder."[3] Rather than focussing exclusively on the actions by, as well as the reactions against, supposed pirates, historians have increasingly turned towards examining piracy as a historical category. Instead of presuming a stable, objective definition of piracy, their work highlights that the accusation of piracy tends to accompany situations in which claims to sovereignty over maritime space are first established, subverted, or openly contested.[4] In this light, the study of the legal and political discourses surrounding and underpinning the charge of piracy can open up questions about power relations, hegemonic ambitions, and permeable jurisdictions that reach far beyond the implications of any specific act of violence at sea.

Writing about piracy "beyond the line", in non-European contexts and settings, is especially challenging because most definitions of piracy are drawn from or directly informed by the historical experience of Europe. This is most obvious in the differentiation that is often drawn between illegal piracy and legal privateering, where the latter denotes an act of piracy that is sanctioned by a state. This distinction ultimately hinges on accepting the authority of a particular state to issue a letter of marque, which turns a criminal pirate into a legitimate privateer.[5] By adopting these categories that are steeped in a specifically European tradition of legal and political thought, historians have inadvertently replicated and affirmed the model of the European nation-state as the implicit standard of political legitimacy.[6] This means that not only

3 Philip Gosse: *History of Piracy*. London: Longmans 1932, p. viii; Michael Pearson: Piracy in Asian Waters. Problems of Definition. In: John Kleinen / Manon Osseweijer (eds): *Pirates, Ports, and Coasts in Asia. Historical and Contemporary Perspectives*. Leiden: IIAS 2010, pp. 15–28, here p. 15.

4 See, for instance, the essays in Robert J. Antony (ed.): *Elusive Pirates, Pervasive Smugglers. Violence and Clandestine Trade in the Greater China Seas*. Hong Kong: HK University Press 2012; as well as the special issue on piracy in Asian waters of *Journal of Early Modern History* 16,6–17,1 (2012–2013).

5 Even Gosse admitted that this distinction could be problematic at times, especially in cases of self-proclaimed privateers bearing commissions "of states which had not yet been born or which did not know of the holders' existence." Gosse: *History of Piracy*, p. viii.

6 On this issue, see Sebastian R. Prange: A Trade of No Dishonor. Piracy, Commerce, and Community in the Western Indian Ocean, Twelfth to Sixteenth Century. In: *American Historical Review* 116,5 (2011), pp. 1269–1293, here p. 1270.

in the imperial archives but also in much of the historiography there exists a strong, inherent bias towards perceiving and representing acts of maritime violence by Europeans as state-sanctioned, and therefore legitimate, while the same acts committed by non-European seafarers are regarded, almost by default, as piracy. As Anne Pérotin-Dumon points out in her pioneering attempt to rethink the history of early modern piracy, most historians have "assumed the legitimacy of European states rather than considering how those states were formed and then established legitimacy over pirates."[7] In studying the history of Europe's maritime empires, it is thus important to question and challenge the category of piracy as selectively applied to acts committed against Europeans or their interests.

This present chapter argues that the construction of piracy in European sources was not a mere corollary but an essential constituent in the projection of European claims to sovereignty over the seas, oceans, and coastlines of the early modern world.[8] What is more, we contend that this piracy discourse has continued to inform historians' treatment of maritime conflicts, which has tended to privilege and legitimize European acts of seaborne larceny. Focusing on the South Asian continent as a central site of European colonial aspirations and contestations in the early modern era, we examine the role and rhetoric of piracy in two colonial encounters: that of the Portuguese on the Malabar Coast in the sixteenth century and that of the British with the Maratha Empire in the eighteenth century. In their combination, these two case studies highlight how the lens, representation, and historiography of European colonialism have defined notions of maritime sovereignty and political legitimacy beyond the line.

2. Lords of Conquest: The Portuguese in the Sixteenth Century

"From the first days of European enterprise in the East, the coasts of India were regarded as a favourable field for filibusters [*i. e.* freebooters], the earliest we hear of being Vincente Sodre [sic], a companion

7 Anne Pérotin-Dumon: The Pirate and the Emperor. Power and the Law on the Seas, 1450–1850. In: C. R. Pennel (ed.): *Bandits at Sea. A Pirates Reader.* New York: NYU Press 2001, pp. 25–54, here p. 28.

8 At a somewhat later date, an analogous process took place on land. See, for instance, Anastasia Piliavsky: The Moghia Menace, or the Watch over Watchmen in British India. In: *Modern Asian Studies* 47,3 (2013), pp. 751–779.

of Vasco da Gama in his second voyage."[9] This opening sentence of *The Pirates of Malabar*, a history of piracy in Indian waters written by the English author John Biddulph a little over a century ago, frankly acknowledges the often piratical nature of European seafaring in the Indian Ocean. The specific reference is to Vincente Sodré, a Portuguese commander in the first decade of the sixteenth century, who was charged with disrupting the Muslim spice trade and who gained notoriety as a vicious predator of native shipping.[10] Sodré is but one example in Biddulph's litany of the deprivations that "English, French, Dutch, Danish, Portuguese, Arab, Malay, or other local pirates" visited upon merchants plying their trade in the Indian Ocean:

> It is difficult to realize, in these days, what a terrible scourge piracy was to the Indian trade [...]. There was no peace on the ocean. The sea was a vast No Man's domain, where every man might take his prey. Law and order stopped short at low-water mark.[11]

From this premise follows a narrative of the English East India Company's efforts to combat piracy, to impose order upon a lawless and violent sea, and to ultimately subdue all challenges to this order by defeating both the French and the Marathas in the eighteenth century. By the time Biddulph published his book, piracy was effectively a matter of the past; in fact, it had become so rare that he intended his book to serve as a reminder to his compatriots of the "tremendous risks and difficulties [...] privations and dangers" that their forefathers had suffered at the hands of pirates whilst securing England's footing in the India trade.[12] *The Pirates of Malabar*, then, proposes a clear contrast between the ravages of the rapacious Portuguese as well as individual European and Asian pirates on the one hand, and the peaceful order established by the English on the other. However, this idea of the East India Company (and, by extension, the British Empire) as a pacifying and law-giving entity was a product of a much later period, originating only in the early nineteenth century with the celebration of

9 John Biddulph: *The Pirates of Malabar; and, An English Woman in India Two Hundred Years Ago*. London: Smith, Elder & Co. 1907, p. 1.

10 Sanjay Subrahmanyam: *The Career and Legend of Vasco da Gama*. Cambridge: Cambridge University Press 1997, esp. pp. 228–234.

11 Biddulph: *The Pirates of Malabar*, pp. vi–vii.

12 Ibid., p. v.

British oceanic hegemony in the wake of the Napoleonic Wars.[13] In the earlier period that Biddulph is describing, it was not the creation of a maritime order but rather the disruption of Iberian hegemonic claims that was at the heart of English seafaring in the Indian Ocean as much as in the Atlantic. Legally excluded from the territories and waterways beyond the line as instituted by papal bull, English mariners enriched themselves (and, when expedient, their monarch) by preying on Spanish and Portuguese shipping. Whether in Francis Drake's letter to his queen (of 1577) that announced his intention to "ruin their sea force" even if he be condemned as a pirate afterwards, or in Walter Ralegh's dictum (ca. 1615) that "whosoever commands the sea commands the trade [...] and consequently the world itself", it is quite apparent that Elizabeth's sea-dogs were, in attitude and practice, self-consciously piratical.[14]

Ironically, at the same time that English pirates were traversing the globe plundering ships and undermining Iberian monopoly claims, English jurists and diplomats were busy trying to uphold the very principle upon which these claims were based. In response to Hugo Grotius propounding the notion of a *mare liberum*, a sea open to all, the English were rallying to the defence of the concept of *mare clausum*, of a "closed sea" subject to the exclusive dominion by a single state. English support for the principle of *mare clausum* represented a major shift in English attitudes from the earlier defiance of the maritime claims advanced by Spain and Portugal. The reason for this about-turn in English policy had to do with the union with Scotland, which had a strong tradition of regarding its waters as a sovereign domain, a rise of more protectionist views in reaction to disputes over coastal fishing and trading rights, as well as the increasingly fierce rivalry with the Dutch over trade with the East Indies where (in marked contrast to Grotius' noble ideas) the Dutch jealously guarded their own monopoly

13 David Armitage: *The Ideological Origins of the British Empire.* Cambridge: Cambridge University Press 2000, p. 100.

14 Letter by Francis Drake dated 6 Nov. 1576, as cited in J. Horton Riley: *Ralph Fitch. England's Pioneer to India and Burma.* London: Fisher Unwin 1899, pp. 31–32; Walter Ralegh: A Discourse of the Invention of the Ships, Anchors, Compass, &c. In: *The Works of Sir Walter Ralegh, Kt*, vol. 8. Oxford: University Press 1829, pp. 317–334, here p. 325.

over the lucrative spice trade.[15] In fact, even though Grotius' treatise had originally been formulated to justify the Dutch capture of a Portuguese ship (the carrack *Santa Catarina*, which was taken in the Straits of Malacca in 1603), the perception at the English court was that "the Dutch put out *Mare Liberum*, made as if aimed at mortifying the Spaniards' usurpation in the W. and E. Indyes, but indeed at England."[16]
Set against this background, it is clear that for the sixteenth and early seventeenth century, Biddulph was right to suggest that "law and order stopped short at low-water mark"[17] in the sense that there was no coherent, international standard of maritime sovereignty against which any universal definition of piracy could be framed. This was true within a single country (such as the shifting attitudes seen in England, or the contradiction between Dutch rhetoric and praxis), within the wider European context (as seen in the disputation of Spanish monopoly claims), but most of all in the clashes between European and non-European conceptions. In these confrontations, the criminalization of indigenous mariners through the accusation of piracy was a key element in the attempt to exclude them from the world of commerce, to marginalize them politically, and to legitimize European "law and order". What is more, this project not only extended to specific merchant groups and seafaring communities but was also applied to entire states whose sovereignty was violated and denied under the pretext of combating piracy.

Portugal was the first European power to enter the Indian Ocean and its king's claim over the ocean as his exclusive domain formed the basis of its official discourse on piracy. This imperial claim was implicit in the title of "Lord of […] the Conquest, Navigation and Commerce of Ethiopia, Arabia, Persia and India" that the Portuguese monarchs adopted from the time of Vasco da Gama's first voyage to India. It was made explicit in the treaties of Alcáçovas (1479) and Tordesillas (1494) and subsequently interpreted by writers such as João de Barros, a chief administrator of Portugal's India trade and a preeminent expositor of

15 Claire Jowitt: *The Culture of Piracy, 1580–1630. English Literature and Seaborne Crime.* Farnham: Ashgate 2012, pp. 3–5.

16 Records prepared for the English embassy to the Congress at Cologne in 1673, as cited in Thomas W. Fulton: *The Sovereignty of the Sea.* Edinburgh: Blackwood 1911, p. 347, fn. 1; cf. Armitage: *Ideological Origins*, pp. 106–119.

17 Biddulph: *The Pirates of Malabar*, pp. vii.

empire, who advanced a legal justification of Portugal's imperial claims that was as elaborate as it was specious. Barros argued that until the arrival of the Portuguese, the Indian Ocean had never been rightfully claimed, with the concept of "rightfully" expressly limited to Christians, thereby at a stroke dismissing all pre-existing claims to sovereignty and suzerainty advanced by Muslim and Hindu rulers.[18] There being no preceding valid title, so Barros, Portugal's king could lawfully claim the ocean as his own domain and restrict access to all others.[19]

His denial of all prior claims to maritime sovereignty, as well as even the basic right to free passage, was the keystone of Portuguese imperialism in Asia. Beginning with Vasco da Gama's second voyage—regarded by Biddulph as the earliest instance of European freebooting in Asian waters—the Portuguese made it their policy not only to restrict indigenous shipping but also to prey relentlessly on their competition in the spice trade. Interpreting his mandate to disrupt the Muslim spice trade in the broadest terms, da Gama systematically attacked and robbed indigenous shipping. A notorious incident from that time is the capture of a merchant ship, identified in Portuguese sources as the *Mīrī*, which was carrying Muslim pilgrims on their return from Mecca to India. Despite the presence of women and children, as well as a number of wealthy merchants who promised immense riches in return for their lives, after plundering its passengers da Gama insisted on burning and sinking the vessel with all 300 souls on board.[20]

Aside from such directly piratical acts, the Portuguese also engaged in more insidious, institutionalized forms of piracy through the imposition of the *cartaz* system. Based on the idea that non-Christians did not enjoy a natural right to free passage, the Portuguese compelled ships traversing the Indian Ocean to purchase safe-passage passes known as *cartazes*. In what was essentially a protection racket, the Portuguese thus used piracy as a threat for extorting indigenous commerce. In practice, not even purchasing a *cartaz* was a guarantor

18 On these earlier claims, see Sebastian R. Prange: The Contested Sea: Regimes of Maritime Violence in the Pre-Modern Indian Ocean. In: *Journal of Early Modern History* 17,1 (2013), pp. 9–33.

19 João de Barros / Diogo do Couto: *Da Ásia*, 24 vols. Lisbon: Regia Officina Typografica 1777, vol. 2, década I, livro viii, cap. 1, pp. 174–175.

20 The different sources for this episode are discussed in Subrahmanyam: *Career and Legend*, pp. 204–210.

of safety, as individual Portuguese captains were often tempted to enrich themselves by plundering merchant vessels, even if it undermined the official racket.[21] Moreover, it was not only private traders who were subjected to this system but indigenous rulers, too. Many coastal potentates obtained passes as part of their treatises with the Portuguese, while others even shared in the profits of selling *cartazes* to Indian, Arab, Persian, and Malay merchants.[22] For instance, in 1513, the Portuguese signed an agreement with the ruler of Calicut, known as the Zamorin ("Çamorim"), that provided for the proceeds from the *cartazes* to be split equally between the kings of Calicut and Portugal.[23] This concession may well have been animated by the key importance that Calicut held in the pepper trade on the Malabar Coast, as well as the hostile relations the Portuguese had had with Calicut, due in no small measure to the intransigence of da Gama. In fact, after having sunk the *Mīrī*, da Gama proceeded to bombard the city for two full days—the Portuguese chronicler described the attack as "an unceasing thunder, and cannonballs of iron and stone rained down, which resulted in great destruction, and in which many people died"—for failing to yield to his demands of expelling all Muslims and paying reparations for a previous conflict.[24]

The alternating states of open enmity and uneasy treaty relations between the Portuguese and the Zamorins of Calicut are a main thread in the history of Portugal's presence in India across the sixteenth century, a history that is not only captured in European sources but also detailed in a contemporary Muslim history of the Malabar Coast. This text, the "Tuhfat al-Mujāhidīn fi ba'd a wāl al Burtughāliyyīn" ("Gift to the Holy Warriors concerning the deeds of the Portuguese", ca. 1583)

21 Michael Pearson: *Coastal Western India: Studies from the Portuguese Records*. New Delhi: Concept 1981, pp. 18–40.

22 K.S. Mathew: Trade in the Indian Ocean and the Portuguese System of Cartazes. In: Malvyn Newitt (ed.): *The First Portuguese Colonial Empire*. Exeter: University of Exeter Press 1986, pp. 69–84, here pp. 73–78.

23 Concerto que D. Garcia de Noronha fez com Çamorim, Rey de Calecut [24 December 1513]. In: J.F. Judice Biker (ed.): *Collecçao de Tratados e concertos de pazes que o Estado da India Portugueza*, vol. II. Lisbon: Imprensa Nacional 1882, pp. 21–23.

24 De Barros / do Couto: *Da Ásia*, vol. 2, década I, livro vi, cap. 5, pp. 52–53. On Kunjali Marakkar, see K.J. John: The Portuguese and Kunjali Marakkar: Myth and Reality. In: Y. Sharma (ed.): *Coastal Histories. Society and Ecology in Pre-Modern India*. Delhi: Primus 2010, pp. 103–116.

by Zayn al-Dīn al-Maʻabarī, is explicit about the Portuguese aggression that was directed primarily against Muslims.[25] It is interesting to note that even though the overarching design of this text is to portray the struggle of Malabar's Muslims against "the Franks", as the Portuguese are referred to, as a legitimate *jihād* (and thus compel other Muslim rulers to assist them), there is a rather frank acknowledgment of the piratical exploits of some elements of the Malabari Muslims.[26] This type of endemic, economically-motivated piracy is contrasted to the systematic persecution of Muslims by the Portuguese, which deprives them not only of their livelihood but also of the freedom with which they had traditionally practiced their religion on the Malabar Coast. Zayn al-Dīn's work devotes a lengthy section to listing the "shameful deeds" of the Portuguese, which include the seizing of Muslim ships, plundering their wealth, and torturing, ransoming, or killing their passengers and crew. And even though it is made explicit that the Portuguese were above all motivated by their ambition to control the valuable pepper trade, there is a pervading sense of religious persecution. In this way, Barros' own religious argumentation for Portugal's right to monopolize the India trade was reflected back in local perceptions of their actions and motivations.

It is not surprising to learn that in some parts of the Indian Ocean world, the Portuguese came to be seen as virtually synonymous to piracy. In Bengal, for instance, they were known as the *harmād*, a corruption of the Portuguese term *armada* (i. e. an armed fleet) which came to serve as a byword for horror and cruelty.[27] As the Portuguese were acquiring a reputation as pirates among traders and coastal communities across coastal Asia, they themselves were liberally deploying the accusation of piracy against their own adversaries. Throughout the sixteenth century, the main antagonists to Portugal's scheme to monopolize the profitable pepper trade were elements of Malabar's Muslim communities. Described in Portuguese sources as *cossários,* these Muslims led attacks on Portuguese shipping, primarily from bases within the kingdom of

25 Zayn al-Dīn al-Maʻabarī: *Tuhfat al-Mujāhidīn fi baʻd a wāl al Burtughāliyyīn.* British Library, India Office Ms. Islamic 2807e.

26 Ibid., fol. 145a.

27 Richard Burn / Wolseley Haig: *The Cambridge History of India. The Mughal Period,* vol. IV. Cambridge: Cambridge University Press 1937, p. 236.

Calicut such as Ponnani. One of their best-known leaders, known as Kunjali Marakkar, was described by the Portuguese as "a pirate [and] enemy of Christians" ("pirata inimigo de Christãos").[28]

However, this label was not only attached to armed opponents: anyone seen as defying Portugal's monopolistic claims as manifested in the *cartaz* system came to be designated a pirate, and treated as such. Kunjali Marakkar for example, who appears as one of the most notorious Muslim pirates ("Mouro cossairo") in the Portuguese records, was part of a Malabari trading community that had thrived as a result of the exodus of foreign merchants from the region under Portuguese duress.[29] However, their harassment at the hand of the Portuguese, most evident in the frequent seizure of their vessels despite being in possession of *cartazes*, alienated the Marakkars to the extent that they eventually abandoned their (in the eyes of the Portuguese) "legitimate" trading ventures to operate outside of the Portuguese system.[30] In the process, they shifted their alliance to the Zamorins of Calicut and became a mainstay of their resistance against the Portuguese in coastal India, eventually even establishing their own thalassocracy on the Malabar Coast.

In the Portuguese sources, native traders became transfigured as interlopers, interlopers as smugglers, and smugglers as pirates. In time, the name of the region itself, *totum pro parte*, came to signify piracy, with the term *"Malavares"* used more or less synonymously for pirates in the official Portuguese records. In this manner, Malabar's history as a key site of resistance to Portugal's imperial designs on the western Indian coast throughout the sixteenth century became recorded as, and reduced to, a story of inveterate piracy. By the time Biddulph penned his book on *The Pirates of Malabar*, he could thus readily present to his audience tales of the "people of the Malabar Coast" as they followed their "hereditary vocation of piracy".[31] In this historiographical tradition, the piracy discourse has served to obfuscate and delegitimize the very real economic, political, and religious rivalries and struggles that

28 De Barros / do Couto: *Da Ásia*, vol. 12, década XII, livro iv, cap. 11, p. 423.

29 De Barros / do Couto: *Da Ásia*, vol. 4, pt. 1, década IV, livro iv, cap. 25, p. 517.

30 See Pius Malekandathil: From Merchant Capitalists to Corsairs: The Muslim Merchants of Malabar and their Responses to Portuguese Maritime Trade Expansion, 1498–1600. In: *Portuguese Studies Review* 12,1 (2004–2005), pp. 75–96.

31 Biddulph: *Pirates of Malabar*, p. 69.

had prompted and fuelled maritime conflicts on the sixteenth-century Indian coast.

3. Piratical States: The British in the Eighteenth Century

The majority of *The Pirates of Malabar* narrates the struggles of the East India Company against the Maratha *sarkhel* (admiral) Angria. To Biddulph, Angria was the "master of the whole coast from Bombay to Vingorla," whose "piracy was a scourge to the European trade of the West coast."[32] Indeed, during the early half of the eighteenth century, the main European and Indian coastal rivalry was between Bombay and the Marathas. Both were seeking to exploit the political disruption left in the wake of Mughal attempts at imperial expansion in the Deccan around the turn of the century, and the successional turmoil that followed the death of Mughal *Padshah* Aurangzeb in 1707.[33] The collapse of Mughal regional authority left the Marathas as the most powerful political and military force in west India and they sought to further extend sovereignty over the maritime spaces along their littoral. At the same time the East India Company endeavoured to eclipse the importance of the Mughal port of Surat, which had entered a period of decline, and establish Bombay as the main trading entrepôt of north-western India.[34] The concurrent rise of both regional powers brought them into almost immediate conflict with each other in which they attempted to extend their jurisdictional boundaries over the region's sea routes and merchant trade. Throughout the conflict the discourse of piracy was invoked by Bombay against Angria, continuing a tradition begun by the Portuguese in the sixteenth century to delegitimize sovereign expressions of state by Indian powers. Eventually the Company defeated Angria in 1756, but only after almost a half-century of near constant warfare on both land and sea. Company historians would remember the victory as one over a "piratical state", and laid

32 Now called Vengurla, the port lies just north of Goa and demarcates the Konkan and Malabar coasts. Biddulph: *Pirates of Malabar*, pp. 76, 77–78.

33 Jadunath Sarkar: *History of Aurangzib Based on Original Sources*, vol. V: The Closing Years, 1689–1707. Calcutta: Sarkar & Sons 1924, p. 236. Ashin Das Gupta: Trade and Politics in Eighteenth Century India. In: Muzaffar Alam / Sanjay Subrahmanyam (eds): *The Mughal State, 1526–1750*. New Delhi: Oxford University Press 2011, pp. 361–397, here p. 364.

34 For the authoritative study of this process see Ashin Das Gupta: *Indian Merchants and the Decline of Surat, c. 1700–1750*. Wiesbaden: Steiner 1994.

the foundations of a British historiographical tradition that Biddulph would later follow.[35]

Largely forgotten outside western India today, Angria's name was virtually synonymous with piracy in eighteenth-century Britain. Newspapers and journals reported on depredations made by Angria against British shipping, which occurred almost every season.[36] For Scottish philosopher David Hume, Angria's name was analogous for a pirate in his 1748 essay "Of the Original Contract".[37] The label of piracy still clings to Angria's name despite colonial and more recent scholarship demonstrating otherwise.[38] For some, confusion lies in the legitimacy of Angria as a subordinate to the Maratha state.[39] Others simply carry forward the historiography of the colonial era.[40] Much of the

35 James Mill: *The History of British India*, vol. 3. 2nd. ed. London: Baldwin, Cradock and Joy 1820, p. 153. Also Robert Orme: *A History of the Military Transactions of the British Nation in Indostan*. London: Nourse 1763, pp. 400–406.

36 For examples see *British Journal*, June 15, 1723, p. 5; *Gentleman's Magazine* 2,20 (1732), pp. 225–928, here p. 927; *London Evening Post*, November 6, 1756, p. 2. Angria was also the subject of the children's book *The Arabian Pirate; or Authentic History and Fighting Adventures of Tulagee Angria*. Newcastle: Angus 1795.

37 Hume wrote: "nor has the greatest and most lawful prince any more reason…to plead a peculiar sacredness or inviolable authority, than an inferior magistrate…or even a robber and a pyrate. The same divine superintendant [sic], who…invested a Titus or a Trajan with authority, did also…bestow power on a Borgia or an Angria." David Hume: Of the Original Contract. In: Eugene F. Miller (ed.): *Essays, Moral, Political, and Literary*. Indianapolis: Liberty Fund Inc. 1987, pp. II.XII.1–II.XII.49, here p. II. XII.3. http://www.econlib.org/library/LFBooks/Hume/hmMPL35.html (accessed May 25, 2013).

38 R. J. Barendse: *The Arabian Seas 1700–1763*, vol. 1. Leiden: Brill 2009, pp. 408–421; A. Deshpande: The Politics and Culture of Early Modern Warfare on the Konkan Coast of India during the Seventeenth and Eighteenth Centuries. In: Sharma (ed.): *Coastal Histories*, pp. 43–73; Derek L. Elliott: *Pirates, Polities and Companies. Global Politics on the Konkan Littoral, c. 1690–1756*. Working Papers No. 136 / 10, Department of Economic History, London School of Economics (2010), pp. 3–6; S. N. Sen: Half a Century of the Maratha Navy. In: *Journal of Indian History* 11,1 (1932), pp. 17–40.

39 Patricia Risso: Cross-Cultural Perceptions of Piracy. Maritime Violence in the Western Indian Ocean and Persian Gulf Region during a Long Eighteenth Century. In: *Journal of World History* 12,2 (2001), pp. 293–319, here pp. 305, 309, 318; Philip Stern: *The Company-State. Corporate Sovereignty and the Early Modern Foundations of the British Empire in India*. New York: Oxford University Press 2011, pp. 186–190; Lakshmi Subramaniam: Piracy in the Indian Ocean. Exploring Perspectives. In: L. Varadarajan (ed.): *Indo-Portuguese Encounters. Journeys in Science, Technology and Culture*, vol. 2. New Delhi: Aryan 2006, pp. 756–763, here p. 761.

40 J. L. Anderson: Piracy and World History. An Economic Perspective on Maritime Predation. In: *Journal of World History* 6,2 (1995), pp. 175–199, here p. 193; Jane Hooper:

confusion and debate amongst scholars stems from the complexities of Maratha history, state formation and Angrian succession. During the period under review, from 1700 to 1756, five different Angrias ruled the Konkan, who were often at odds with each other and the Maratha *Chhatrapati* (paramount sovereign) or *Peshwa* (prime minister). The British employed the discourse of piracy against each of them, though to varying degrees, depending upon the diplomatic objectives and political constraints facing Bombay. Thus during the early eighteenth century the East India Company adopted a rather contradictory position that could simultaneously recognise the sovereignty of Angria, yet delegitimise the methods that were used to exercise that authority over the coast. The result was a balancing act between strengthening Bombay's increasing regional commercial hegemony and remaining on favourable terms with the larger and more powerful Maratha land-based empire.

The Angria family first rose to prominence when Maratha *Chhatrapati* Sambhaji granted Kanhoji Angria with a hereditary admiralship (*sarkhel*) in 1688.[41] The title gave Kanhoji control of the island fortress of Survarnadurg and placed him in charge of the Maratha naval force. Kanhoji soon gained control of the entire Konkan coast and built up the Maratha fleet. He also instituted the *dastak*, a system similar to the Portuguese *cartaz*, by which all ships that sailed his waters and called at his ports were required to purchase his pass. Kanhoji died in 1729 leaving his *sarkhel* and territory to his son Sukhoji who ruled the coast in similar fashion for four short years until his death in 1733. Succession was then contested by two of Kanhoji's other sons, Sambhoji and Manaji. Eventually, the Maratha *Peshwa* Baji Rao I mediated the dispute granting Manaji control over the northern Konkan, and Sambhoji over the south.[42] Bombay also involved itself in the fraternal strife seeing it as, "a good Opportunity to endeavour to weaken our Enemies force by keeping up the dissention between the two Brothers", and backed

Pirates and Kings. Power on the Shores of Early Modern Madagascar and the Indian Ocean. In: *Journal of World History* 22,2 (2011), pp. 215–242, here p. 225.

41 Manohar Malgonkar: *Kanhoji Angrey, Maratha Admiral. Account of His Life and Battles with the English*. Bombay: Asia Publishing House 1959, pp. 54–55.

42 André Wink: *Land and Sovereignty in India. Agrarian Society and Politics under the Eighteenth-Century Maratha Svarajya*. Cambridge: Cambridge University Press 1986, p. 76.

the weaker Manaji.[43] Until his death in 1759, Manaji and the Company maintained an uneasy alliance that allowed vessels under Bombay's protection to transit freely through the region without requiring the purchase of his *dastak*.[44] Meanwhile in the south, Sambhoji Angria possessed a larger navy and was thus better able to resist Bombay's coercive efforts, a fact made clear in 1735 when he captured the richest prize in Angrian history, the *Derby*, carrying away £32,000 in cash, the crew and other valuable cargo.[45] Tulaji Angria succeeded Sambhoji as lord of the south in 1743 and continued his brother's (and father's) policy of demanding *dastaks* from all vessels sailing his waters, including the Company's.

The British began referring to the Angrias as pirates in the earliest years of the eighteenth century. In a series of instructions to Company servant William Reynolds, sent to enter into a treaty with Angria, Kanhoji is called "a Savajee Pirate independent on that Raja,"[46] who cannot "be permitted searching Molesting or Seizing any […] Boates, or other Vessells, from what Port, Harbour, Place of what Nation so ever they maybe, bringing Provisions, Timber, or Merchandize to Bombay".[47] Two important issues are raised here. The first is indicative of the East India Company's peripheral position in western India. Bombay was not completely aware of the political situation on the mainland. Isolated geographically, linguistically, culturally and politically, Bombay either misunderstood, or was ignorant of the conditions and disarray within the Maratha Empire at the time, a point made clear in their description of Kanhoji as *both* a Maratha and independent of the *Chhatrapati*. Decades-long warfare with the Mughals, and the untimely death of *Chhatrapati* Rajaram had left the state barely functioning, and as a result,

43 Maharashtra State Archives (MSA). Bombay Public Department Diary (BPDD), vol. 7. Consultation. November 22, 1734.

44 MSA. Marine Superintendent Office Diary (MSOD), vol. 4. Entry for December 23, 1742. For more on Manaji Angria and his successors in the north Konkan, see Simon Layton: The "Moghul's Admiral". Angrian "Piracy" and the Rise of British Bombay. In: *Journal of Early Modern History* 17,1 (2013), pp. 75–93, here pp. 87–93.

45 *A Faithful Narrative of the Capture of the Ship Derby*. London: Osborn 1738, pp. 55–56.

46 "Savajee" was a British appellative for the Marathas – a corruption of Shivaji, the name of the empire's founder. British Library (BL). India Office Records (IOR)/P/341/2. Consultation. November 24, 1704, p. 15. Also cited in Stern: *The Company-State*, p. 187.

47 BL.IOR/P/341/2. Consultation. November 24, 1704, p. 17.

its regional governors more-or-less autonomous.[48] However, despite the near disintegration of centralised authority, Kanhoji never acted nor claimed to be independent of the Maratha government. Responding to Reynolds' treaty terms, Kanhoji wrote:

> the Sevajees [...] had held out Warr with the Mogull forty years; lived now by their Sword and would seize what Boates or other Vessell belonging either to the Mogulls Vessells from any of his Ports, or Mallabarr excepting such as had Conjee Angras Passports; the English being at liberty Acting as they please.[49]

By asserting his mandate under and in the name of the Maratha sovereign he claimed his authority to enforce his *dastak* upon non-aligned and enemy ships and traders. Angria was unwilling to allow the English to dictate the terms of Indian politics, but recognised their right to act "as they please" and demand passes from whomever they desired. To Kanhoji, this was a justifiable expression of Bombay's sovereignty, just as it was his. By 1708, Bombay had come to realise that Angria was not independent and acted under a suzerain authority. Several times they appealed directly to the Maratha ruler to bring Kanhoji to the favourable terms they had originally proposed in 1704, leading to occasional periods of uneasy peace.[50] Ultimately amity between the two would breakdown over interpretations of jurisdiction, and when it did, Kanhoji again became a "Sevajee Pyratt" but was no longer independent.[51]

This raises the interrelated second issue, which is the deployment of the pirate discourse to delegitimize Angria's expression of authority over the coast. It also reveals Bombay's principle political aim. At the time of Reynolds' offer of terms, Bombay and Angria had a pre-existing agreement that they would not seize vessels belonging to each other. Bombay interpreted this as meaning any ship sailing for the Company, whether it came from Europe, Madras, or Bombay itself. They further considered all Indian country-trading vessels bound to Bombay as falling under Company jurisdiction. Angria on the other hand, only recognised East India Company ships carrying Bombay's

48 Stewart Gordon: *The Marathas 1600–1818. The New Cambridge History of India*, vol. II.4. Cambridge: Cambridge University Press 2006, p. 98.

49 BL.IOR/P/341/2. Consultation. December 2, 1704, p. 30.

50 BL.IOR/P/341/3. Bombay to Sou Raja. January 29, 1708, n. p.

51 BL.IOR/P/341/3. Consultation. September 8, 1707, n. p.

pass as belonging to the island. To Angria, Indian and British vessels under any authority other than Bombay also had to purchase his *dastak* to ensure safe conduct.[52] This was an unacceptable position for Bombay, which refused to purchase the passes of any nation, Indian or European.[53] Due to Bombay's inability to enforce terms favourable to the island's trade and political position in Western India, it relied on the discourse of piracy to delegitimize Maratha jurisdiction over maritime spaces–despite the fact that Bombay recognised the sovereignty of the Maratha state and Angria's position within it. This was a strategy used against all the Angrias whenever they could not be coerced into accepting the Company's regional hegemony. Manaji proved the only exception, due to his weaker position relative to his more powerful southern brothers with whom he was in enmity, which necessitated his compliance with the Company's maritime regime.

The primary concern for the governors of Bombay at this time was increasing the island's importance to the Indian trade. This guided their policy of extending sovereignty over European and Indian country-trading vessels coming to and emanating from the island. Bombay feared that the increased transaction costs to Indian merchants as a result of having to purchase Angria's *dastak* would cause the Company to "absolutely loose this Incourageing Juncture bringing the Indostan Trade down to [this] Island".[54] Through governor John Child's disastrous war with the Mughals in the late seventeenth century, Bombay had learned that it was unable enforce its objectives militarily on its Indian neighbours.[55] Instead, the Company used its pass system to attract trade and encourage settlement to Bombay. Regular convoys were established that escorted ships along the coastal trade routes protecting them from Angria's cruisers. It was a strategy that was largely successful, even if it meant increased costs to Bombay and the occasional loss of vessels. Security of trade mutually reinforced the two

52 For more on the regional politics of the pass-system, see Derek L. Elliott: The Politics of Capture in the Eastern Arabian Sea, c. 1700–1750. In: *International Journal of Maritime History* 25,2 (2013), pp. 187–198.

53 BL.IOR/P/341/2, Reply to Portuguese, January 30, 1706, p. 152; MSA.BPDD, vol. 2,2. Richard Bourchier to Tulaji Angre, October 30, 1754, n. p.

54 BL.IOR/P/341/2. Consultation. October 30, 1705, p. 70.

55 John Richards: *The Mughal Empire. The New Cambridge History of India*, vol. I.5. New Delhi: Cambridge University Press 1993, pp. 239–240.

contradictory principles of establishing a British *mare clausum* through the enforcement of its pass, and encouraging free trade (under British commercial hegemony) to Bombay.

Lakshmi Subramaniam has argued that the type of politics enforced through the Company's pass was for all practical purposes the same as Angria's, with the only distinction being that they did so under the rhetoric of protecting "the ordinary trader."[56] *Mare liberum* for country-traders was supported by the Company, but only if under Bombay's control. To other regional powers the trade routes were closed unless they accepted the regional hegemony of Bombay. This justification of establishing British law and order on the seas was concomitant with the guiding political economic ethos of the day. Freedom of trade ensured through a lawful–and thus legitimate–security apparatus was viewed as the most effective way to establish Bombay as the most important regional centre of trade.[57] Kanhoji Angria and his sons (with the usual exception of Manaji) stood in the way of the Company's designs by refusing to concede to British attempts at maritime and commercial hegemony. As such, Angrian political claims over the littoral and enforcement of their sovereignty through the *dastak* were considered illegitimate, and thus the actions of a "piratical state".

Perhaps the aspect most confusing to scholars in the historiography of the Angrias concerns Tulaji, who took control over the south Konkan in 1743 and ruled until his military defeat by an allied Anglo-Maratha force in 1756. On the surface, the fact that the Marathas attacked him seems to give credence to the pirate narrative established by the British. However this again obfuscates and ignores particular political developments and divisions within the Maratha Empire unknown to Bombay. The first years of Tulaji's rule were no different from his forbearers Sambhoji, Sukhoji and Kanhoji. He governed the south Konkan continuing to exercise Maratha influence on both land and sea. He demanded that all vessels traversing his waterways carry his *dastak* and captured country and British ships that failed to do so. Relations between Tulaji and his Maratha masters began amicably and in 1744, *Peshwa* Balaji Baji Rao even asked Bombay to put "aside your

56 Lakshmi Subramaniam: Bombay and the West Coast in the 1740's. In: *Indian Economic and Social History Review* 18,2 (1981), pp. 189–216, here p. 193.

57 Stern: *The Company State*, pp. 84–86.

disagreement with Tulajee and do not give any molestation to his Fleet" so he could sail northward unhindered, in order to make a coordinated attack on a Maratha enemy.[58]

However by the end of the decade relations between Tulaji and the state became strained. Records are unclear as to the exact nature of the rift, but by the early 1750s Tulaji was reportedly oppressing the Brahmin population of the Konkan, invading the territories of allied neighbours in order to increase his possessions, and failing to remit revenue to Maratha centralised authority.[59] These rebellious actions took place following the death of *Chhatrapati* Shahu in 1749, and coincided with the usurpation of the sovereign's powers by the *Peshwa*, which placed the control of the Maratha state fully into the prime minister's hands.[60] This was a controversial manoeuvre that was not supported unanimously by the Maratha land-holding nobility, many of whom wanted power continued with the *Chhatrapati*. According to Grant Duff, the foremost colonial-era historian of the Maratha Empire, Tulaji was one of these and, "disavowed the Peishwa's authority".[61] This explains his sudden belligerency towards the state, not so much as a bid for independent power, but as refusing to submit to the control of the *Peshwa*. Several warnings were sent to Tulaji to get him to desist, but ultimately unable to bring him under control, the *Peshwa* wrote to Bombay in 1755 requesting their assistance in removing him from power.[62]

If any Angria can be considered a pirate at all, in the sense that they acted without the sanction of state authority, it is Tulaji. However, the politically illegitimate actions that he took were not against the British, but against the Marathas–his own state. For the *Peshwa*, Tulaji's enforcement of his *dastak* on the seas against growing British hegemony was a valid expression of his political authority. It was only when he attempted to extend his own power within the Maratha Empire

58 MSA. BPDD, vol. 17,2. Translate of Letter from Ballajee Badjerow, December 17, 1744, n. p.

59 *Selections from the Satara Raja's and the Peishwa's Diaries*, vol. 1, ed. by Rao Bahadur Ganesh Chimnaji Vad. Satara: Sahasrakar Press 1902, pp. 111–116.

60 Richard Eaton: *A Social History of the Deccan, 1300–1761. Eight Indian Lives. The New Cambridge History of India*, vol. I.8. Cambridge: Cambridge University Press 2005, p. 195.

61 James Grant Duff: *History of the Mahrattas*, vol. 2. Bombay: Exchange Press 1863, p. 59. For the *peshwa* rise to power, see ibid., pp. 21–30.

62 MSA. Secret and Political Department Diary (S&PDD), vol. 1. Consultation. March 10, 1755. n. p.

that Tulaji overstepped his prerogative. After their victory the *Peshwa* wrote to Bombay that "Toolajee Angria was a servant to this state, but for want of sense [...] turned aside from obedience and begun to be a Rebel and proud, it was therefore thought necessary to punish and subdue him".[63] For the British though, this would not matter. Tulaji's actions at sea against the Company's trade were "piratical" in that they ran contrary to British projections of pre-colonial power. As a result, for contemporaries, and later historians such as Biddulph, the legacy of Angria would be rulers of a "piratical state".

4. "The splendid names of things"

In his narrative of the encounter between Alexander the Great and the captured pirate, St Augustine made no secret of his own attitude by declaring the pirate's bold retort an answer both "pertinent and true".[64] What, Augustine asks, truly differentiates the pirate from the emperor when both are governing groups of men who are bound together by a compact of association (*pactum societatis*) and who divide the spoils of their enterprise according to an agreed-upon law?

> If, by the constant addition of desperate men, this scourge grows to such a size that it acquires territory, establishes a seat of government, occupies cities and subjugates people, it assumes the name of kingdom more openly. For this name is now manifestly conferred upon it not by the removal of greed, but by the addition of impunity.[65]

In retelling the history of empire—in *City of God*, this being above all the history of Rome—Augustine cautions his readers to remain wary of the presumption of legitimacy prompted by longevity, scale, or glory. He urges not to let critical faculties "be dulled by the splendid names of things when we hear of 'peoples', 'kingdoms' and 'provinces'."[66] By divesting empire of its claim to natural legitimacy, Augustine makes it possible for even the most paradigmatic of outlaws, the piratical *hostis humanis generis*, to be conceived of as an equal to the emperor, for "what are bands of robbers but little kingdoms?"[67]

63 S&PDD, vol. 3. Letter from Nannah, February 12, 1757. In: B. K. Shrivastavya (ed.): *The Angreys of Kolaba in British Records*. Poona: Prashant Printery 1950, p. 150.

64 Augustine: *The City of God*, p. 147.

65 Ibid.

66 Ibid, p. 146.

67 Ibid, p. 147.

To Augustine, the only meaningful distinction between pirates and empires could not be a long history, or sheer size, or grand titles, but justice. Inverting his analogy of robber bands as little kingdoms, Augustine argues that when the principle of justice is removed, those who hold political power are nothing more than "great bands of robbers".[68] By taking Augustine's analogy as a point of departure, the discourses of piracy in Europe's imperial encounter with India is revealed as both fluid and contested. In the archive and historiography, its conception has been dominated by the perspectives and categories of those who ultimately came to control India and its trade. It was through the enforcement of European normative conceptions of justice and law that Indian states were left devoid of legitimacy and thus reduced to the status of "pirate" both politically as well as morally. In the maritime spaces of western India, the overlapping jurisdictions of European and Indian claims and domains were hotly contested but ultimately settled by European naval power.[69] Justice for the Portuguese in the sixteenth century was based on a religious imperative through which they claimed a monopoly in trade and denied the right of free passage to non-Christians. Two centuries later, British justice took the form of defending the rights of Indian merchants under the rubric of free trade against the Maratha *sarkhels.*

Both European powers sought to establish hegemonic claims to the waters of the Malabar Coast. The relative success of each can be seen in the types of discourse levelled against them by Indians. Over the course of the sixteenth century, Portuguese legitimacy was tarnished by their actions and disputed by Indian communities who fought against them. It was also challenged by other European powers eager to disrupt and undermine their monopoly claims by appealing to the notion of a *mare liberum.* The British, on the other hand, defended—even if they did so reluctantly and under Mughal coercion—Surati shipping against both European and Indian pirates at the turn of the eighteenth century.[70] Though the British acted in ways that were not all

68 Augustine: *The City of God.*

69 On the role of European sailing technology, see Tonio Andrade: Was the European Sailing Ship a Key Technology of European Expansion? Evidence from East Asia. In: *International Journal of Maritime History* 23,2 (2011), pp. 17–40.

70 Stern: *The Company-State*, pp. 134–137.

that dissimilar to their Lusitanian predecessors, they were able to avoid being labelled as pirates, even by their enemies. One crucial difference is that within decades of the British defeat of Angria they went on to become the major territorial power in India, which in an Augustinian trajectory may be seen as the addition of impunity to an inviolate greed.

This chapter demonstrates how the category of piracy was applied to delegitimize the actions of Indian states that challenged the hegemonic claims of Europeans in the early modern period. In drawing attention to how the colonial category of piracy was constructed and later adopted, often uncritically, by historians such as Biddulph, we echo Kären Wigen's call to approach maritime history "as an analytical project [...] extending not only from the ships to the docks [...] [but] including littorals and their hinterlands as well."[71] Connecting the history of the sea to land-based events as well as developments in political thought and theory allows for a more complete picture that helps to deconstruct notions of illegitimacy and piracy that have distorted interpretations of European as well as Indian actions. So the question that this history of maritime conflict and contestation in Europe's colonial encounters with India prompts is not so much "who is a pirate" but how the underlying contradictions—between disorder and hegemony, between the notions of *mare clausum* and *mare liberum*, between trade and violence, between claims and counter-claims—came to be addressed, contested, and resolved over the course of the early modern era. Put in other words, the question is how are we to write the history of maritime imperialism beyond the line: purely as the story of the emperor, or also as that of the pirate?

71 Kären Wigen: Introduction. In: J. H. Bentley / R. Bridenthal / K. Wigen (eds): *Seascapes. Maritime Histories, Littoral Cultures, and Transoceanic Exchanges.* Honolulu: University of Hawaii Press 2007, pp. 1–18, here pp. 12.

Bibliography

Anderson, J. L.: Piracy and World History. An Economic Perspective on Maritime Predation. In: *Journal of World History* 6,2 (1995), pp. 175–199.

Andrade, Tonio: Was the European Sailing Ship a Key Technology of European Expansion? Evidence from East Asia. In: *International Journal of Maritime History* 23,2 (2011), pp. 17–40.

Antony, Robert J. (ed.): *Elusive Pirates, Pervasive Smugglers. Violence and Clandestine Trade in the Greater China Seas.* Hong Kong: HK University Press 2012.

Antony, Robert J. / Sebastian R. Prange (eds): Special Issue: Piracy in Asian Waters. *Journal of Early Modern History* 16,6–17,1 (2012–2013).

The Arabian Pirate; or Authentic History and Fighting Adventures of Tulagee Angria. Newcastle: Angus 1795.

Armitage, David: *The Ideological Origins of the British Empire.* Cambridge: Cambridge University Press 2000.

Augustine, Bishop of Hippo: *The City of God against the Pagans,* ed. by R.W. Dyson. Cambridge: Cambridge University Press 1998.

Barendse, R.J.: *The Arabian Seas 1700–1763*, vol. 1. Leiden: Brill 2009.

De Barros, João / Diogo do Couto: *Da Ásia.* 24 vols. Lisbon: Regia Officina Typografica 1777–1788.

Biddulph, John: *The Pirates of Malabar; and, An Englishwoman in India Two Hundred Years Ago.* London: Smith, Elder & Co. 1907.

Biker, J. F. Judice (ed.): Concerto que D. Garcia de Noronha fez com Çamorim, Rey de Calecut. In: *Collecçao de Tratados e concertos de pazes que o Estado da India Portugueza*, vol. II. Lisbon: Imprensa Nacional 1882, pp. 21–23.

British Journal, June 15, 1723, p. 5.

British Library, India Office Records: P Series, Range 341, vols. 2–3. London, UK.

Burn, Richard / Wolseley Haig: *The Cambridge History of India. The Mughal Period*, vol. IV. Cambridge: Cambridge University Press 1937.

Chomsky, Noam: *Pirates and Emperors, Old and New. International Terrorism in the Real World.* Rev. ed. Cambridge, MA: South End Press 2002.

Cicero, Marcus Tullius: *Cicero's Tusculan Disputations. Also, Treatises on the Nature of the Gods, and on the Commonwealth*, trans. from the Latin by C. D. Yonge. New York: Harper 1877.

Das Gupta, Ashin: Trade and Politics in Eighteenth Century India. In: Muzaffar Alam / Sanjay Subrahmanyam (eds): *The Mughal State, 1526–1750.* New Delhi: Oxford University Press 2011, pp. 361–397.

—: *Indian Merchants and the Decline of Surat, c. 1700–1750.* Wiesbaden: Steiner 1994.

Deshpande, Anirudh: The Politics and Culture of Early Modern Warfare on the Konkan Coast of India during the Seventeenth and Eighteenth Centuries. In: Y. Sharma (ed.): *Coastal Histories. Society and Ecology in Pre-Modern India.* Delhi: Primus 2010, pp. 43–73.

Duff, James Grant: *History of the Mahrattas*, vol. 2. Bombay: Exchange Press 1863.

Eaton, Richard: *A Social History of the Deccan, 1300–1761. Eight Indian Lives. The New Cambridge History of India*, vol. I.8. Cambridge: Cambridge University Press 2005.

Elliott, Derek L.: The Politics of Capture in the Eastern Arabian Sea, c. 1700–1750. In: *International Journal of Maritime History* 25,2 (2013), pp. 187–198.

—: *Pirates, Polities and Companies. Global Politics on the Konjan Littoral, c. 1690–1756.* Working Papers No. 136 / 10, Department of Economic History, London School of Economics (2010).

A Faithful Narrative of the Capture of the Ship Derby. London: Osborn 1738.

Fulton, Thomas W.: *The Sovereignty of the Sea.* Edinburgh: Blackwood 1911.

Ganesh Chimnaji Vad, Rao Bahadur (ed.): *Selections from the Satara Raja's and the Peishwa's Diaries*, vol. 1. Satara: Sahasrakar Press 1902.

Gentleman's Magazine 2,20 (1732), pp. 225–928.

Gordon, Stewart: *The Marathas 1600–1818. The New Cambridge History of India*, vol. II.4. Cambridge: Cambridge University Press 2006.

Gosse, Philip: *History of Piracy.* London: Longmans 1932.

Harding, Brian: The Use of Alexander the Great in Augustine's *City of God.* In: *Augustinian Studies* 39,1 (2008), pp. 113–128.

Hooper, Jane: Pirates and Kings. Power on the Shores of Early Modern Madagascar and the Indian Ocean. In: *Journal of World History* 22,2 (2011), pp. 215–242.

Hume, David: Of the Original Contract. In: Eugene F. Miller (ed.): *Essays, Moral, Political, and Literary.* Indianapolis: Liberty Fund Inc. 1987, pp. II.XII.1–II.XII.49. http://www.econlib.org/library/LFBooks/Hume/hmMPL35.html (accessed May 21, 2013).

John, K. J.: The Portuguese and Kunjali Marakkar: Myth and Reality. In: Y. Sharma (ed.): *Coastal Histories. Society and Ecology in Pre-Modern India.* Delhi: Primus 2010, pp. 103–116.

Jowitt, Claire: *The Culture of Piracy, 1580–1630. English Literature and Seaborne Crime.* Farnham: Ashgate 2012.

Layton, Simon: The "Moghul's Admiral". Angrian "Piracy" and the Rise of British Bombay. In: *Journal of Early Modern History* 17,1 (2013), pp. 75–93.

London Evening Post, November 11, 1756, p. 2.

Maharashtra State Archives. Bombay Public Department Diaries, vols. 7 and 17,2. Mumbai, India.

Maharashtra State Archives. Marine Superintendent Office Diary, vol. 4. Mumbai, India.

Maharashtra State Archives. Secret and Political Diaries, vols. 1 and 3. Mumbai, India.

Al-Maʿabarī, Zayn al-Dīn: *Tuhfat al-Mujāhidīn fi ba'd a wāl al Burtughāliyyīn*. British Library, India Office Ms. Islamic 2807e.

Malekandathil, Pius: From Merchant Capitalists to Corsairs: The Muslim Merchants of Malabar and their Responses to Portuguese Maritime Trade Expansion, 1498–1600. In: *Portuguese Studies Review* 12,1 (2004–2005), pp. 75–96.

Malgonkar, Manohar: *Kanhoji Angrey, Maratha Admiral. Account of his Life and Battles with the English*. Bombay: Asia Publishing House 1959.

Mathew, K. S.: Trade in the Indian Ocean and the Portuguese System of Cartazes. In: Malvyn Newitt (ed.): *The First Portuguese Colonial Empire*. Exeter: University of Exeter Press 1986, pp. 69–84.

Mill, James: *The History of British India*, vol. 3. 2nd. ed. London: Baldwin, Cradock and Joy 1820.

Orme, Robert: *A History of the Military Transactions of the British Nation in Indostan*. London: Nourse 1763.

Pearson, Michael: Piracy in Asian Waters. Problems of Definition. In: John Kleinen / Manon Osseweijer (eds): *Pirates, Ports, and Coasts in Asia. Historical and Contemporary Perspectives*. Leiden: IIAS 2010, pp. 15–28.

—: *Coastal Western India: Studies from the Portuguese Records*. New Delhi: Concept 1981.

Pérotin-Dumon, Anne: The Pirate and the Emperor. Power and the Law on the Seas, 1450–1850. In: *Bandits at Sea. A Pirates Reader*, ed. by C. R. Pennel. New York: NYU Press 2001, pp. 25–54.

Piliavsky, Anastasia: The Moghia Menace, or the Watch over Watchmen in British India. In: *Modern Asian Studies* 47,3 (2013), pp. 751–779.

Prange, Sebastian R.: The Contested Sea. Regimes of Maritime Violence in the Pre-Modern Indian Ocean. In: *Journal of Early Modern History* 17,1 (2013), pp. 9–33.

—: A Trade of No Dishonor. Piracy, Commerce, and Community in the Western Indian Ocean, Twelfth to Sixteenth Century. In: *American Historical Review* 116,5 (2011), pp. 1269–1293.

Ralegh, Walter: A Discourse of the Invention of the Ships, Anchors, Compass, &c. In: *The Works of Sir Walter Ralegh, Kt*, vol. 8. Oxford: University Press 1829, pp. 317–334.

Richards, John: *The Mughal Empire. The New Cambridge History of India*, vol. I.5. New Delhi: Cambridge University Press 1993.

Riley, J. Horton: *Ralph Fitch. England's Pioneer to India and Burma*. London: Fisher Unwin 1899.

Risso, Patricia: Cross-Cultural Perceptions of Piracy. Maritime Violence in the Western Indian Ocean and Persian Gulf Region during a Long Eighteenth Century. In: *Journal of World History* 12,2 (2001), pp. 293–319.

Sarkar, Jadunath: *History of Aurangzib Based on Original Sources*, vol. 5: The Closing Years, 1689–1707. Calcutta: Sarkar & Sons 1924.

Sen, S. N.: Half a Century of the Maratha Navy. In: *Journal of Indian History* 11,1 (1932), pp. 17–40.

Shrivastavya, B. K. (ed.): *Angreys of Kolaba in British Records*. Poona: Prashant Printery 1950.

Stern, Philip: *The Company-State. Corporate Sovereignty and the Early Modern Foundations of the British Empire in India*. New York: Oxford University Press 2011.

Subramaniam, Lakshmi: Bombay and the West Coast in the 1740's. In: *Indian Economic and Social History Review* 18,2 (1981), pp. 189–216.

—: Piracy in the Indian Ocean. Exploring Perspectives. In: L. Varadarajan (ed.): *Indo-Portuguese Encounters. Journeys in Science, Technology and Culture*, vol. II. New Delhi: Aryan 2006, pp. 756–763.

Subrahmanyam, Sanjay: *The Career and Legend of Vasco da Gama*. Cambridge: Cambridge University Press 1997.

Wigen, Kären: Introduction. In: J. H. Bentley / R. Bridenthal / K. Wigen (eds): *Seascapes. Maritime Histories, Littoral Cultures, and Transoceanic Exchanges*. Honolulu: University of Hawaii Press 2007, pp. 1–18.

Wink, André: *Land and Sovereignty in India. Agrarian Society and Politics under the Eighteenth-Century Maratha Svarajya*. Cambridge: Cambridge University Press 1986.

Swaraj for Kenya, 1949–1965

The Ambiguities of Transnational Politics *

Margret Frenz

> Nationalism has blinded us to the possibility of connection, and historical ethnography, whether in one of its western variants of high Orientalism, or whether practised in the East, has aided and abetted this unfortunate process.[1]

Introduction

Anti-imperial movements across the global south were transnational—indeed, transimperial—undertakings. Connections across the Indian Ocean existed before the British set up their empire along its shores, but they were transformed to suit the new rulers during the nineteenth and early twentieth centuries. Transimperial anti-colonialism mirrored the entangled histories of Empire. At the same time, transoceanic connections proved to be vital for the liberation movement in British East Africa, particularly Kenya. In addition, the process of decolonization, the emergence of nation-states in South and South East Asia and the intention of the British and the French to reconstruct their Empires—at least in Africa—had major impacts on these transimperial / transnational connections. Analysing these connections and trajectories from a south-south perspective provides an enriched historical understanding of the emergence of a post-1945 world order.

* This essay was first published in *Past & Present* 218 (2013), pp. 151–177. We wish to thank the publisher for the permission to reproduce Margret Frenz' essay in this book.

1 Sanjay Subrahmanyam: Connected Histories: Notes towards a Reconfiguration of Early Modern Eurasia. In: *Modern Asian Studies* 31,3 (1997), pp. 735–762, here p. 761.

Transimperial and transnational connections can be understood in terms of different phases and places. One is the rise and then fall of transimperial nationalist movements. Secondly, transformation processes evolved and were shaped differently in various locations such as Kenya, Tanzania, and Ghana. Roughly speaking, the rise of transimperial / transnational connections began in the pre-World War I era, intensified over the inter-war years, dropped during World War II and experienced a steep rise after 1945. But with the advent of Kenyan independence in 1963 and the process of building a new nation-state, these transnational connections played less and less of a role both within and outside the Kenyan body politic.

The connected, entangled transimperial / transnational history of nationalism across the Indian Ocean is the focus of this article. It reflects the intellectual, economic, and personal links between members of the South Asian diaspora in Kenya, African Kenyans and Indians, during the decades leading to independence in India in 1947 and in Kenya in 1963. These underwent different phases—of intense and less intense contact. The connections provided through the South Asian diaspora were a constituent part of the Kenyan nationalist movement, reflecting a south-south transimperialism and transnationalism. Fundamental to the analysis of south-south connections is the realization that nationalism not only involved communicating ideas of *swaraj* / *uhuru* and self-determination, but also required a practice and material substance in order to achieve its goals. In this article I argue that for the latter transimperial / transnational connections were of the utmost importance. While African nationalists in the pre-independence period drew heavily on transimperial / transnational connections to further their cause, powerful 'nation-builders' in the post-independence period saw such connections as threats that might undermine both their own legitimacy and their country's national sovereignty.

The opaque and changing relationship between Asian and African communities in Kenya helps to explain the ambivalences at the heart of recognizing the roles played by Joseph Murumbi—the first Kenyan Foreign Minister and Second Vice-President—and other Kenyan Indians.[2] Several public intellectuals in Kenya have started to explore the

2 Note on terminology: Until the independence of India and Pakistan, individuals from the sub-continent were mostly referred to as 'Indians' or as 'Goans' in East

complex and intricate participation of a few, but nonetheless prominent and influential, Kenyan South Asians in the nationalist movement.[3] Apparently, a shift from an inward-looking nationalism towards an outward-looking nationalism is under way, a shift that, at least to some extent, acknowledges the different voices and the cooperation and negotiation that characterized the anti-colonial movement in the 1950s and 1960s. A similar phenomenon can be observed in academic publications, where the role of South Asians in East African politics is finally being critically examined.

A transnational flow and exchange of ideas depends on individual actors who transport those ideas in a wide range of settings, crossing existing boundaries and acting as social, political, or economic entrepreneurs. Dimensions for an analysis of such a global flow of ideas are firstly, a physical space in which such a phenomenon occurs: in my case study, the interfaces between the Indian Ocean region and the political entity of the British Empire. A second dimension is that of a particular intellectual framework to provide a basis for debate, one that proves to be part of a specific *Zeitgeist* that is attractive for a large number of people, for instance nationalism or communism. A third dimension is a favourable historical context—often connected with a major 'disruptive' event—which opens up new possibilities. In the present case, the gradual demise of the British Empire in the wake of the Second World War triggered the re-arrangement of political entities along the Indian Ocean rim that was part of the emerging post-1945 world order. Fourthly, ideas do not travel in an institutional vacuum: without political parties, the involvement of mass media, or social movements, they disappear from view with few traces for the historian to salvage. In other words, the transimperial and transnational circulation of ideas is bound to people and their realization in institutions large enough to implement them.

Indian and Pakistani independence in 1947 and the immediate post-colonial years had a much larger impact on politicians and politics in

Africa. Thereafter, the term 'Asians' was increasingly used in order to include individuals hailing from the new nation-states. Opinion and usage of terms remain controversial, however, but the most widely used today is 'Asians'.

3 See, e.g., *The Making of a Nation. A Political History of Kenya* (Kenya, 2008 D: Hilary Ngweno).

East Africa than has hitherto been assumed.[4] The transnational movement of ideas and individuals in the turbulent period between the end of the Second World War and the early post-colonial years was crucial to the nationalist movement in East Africa. Recognizing these ideas, networks, individuals and resources is essential to a fuller understanding of the connected histories of South Asia and East Africa, and the transformation of their relationships with the metropolis.

In this paper, I analyse first the shift from official silence to recognition. I then reflect on the deeper message: how to understand the roles of individuals who carry transnational flows of ideas and influence, and the roles of social institutions such as political parties that then embody or act on such ideas in public spaces. In the final part of this paper, I return to the issue of the historical significance of transimperial / transnational connections for the analysis of nationalist movements.

Public and academic debate

In current historiographical writing, reconstructions and interpretations of the nationalist movement tend to focus on the conflict between African Kenyans and Europeans, with the role of Indian Kenyans in the anti-colonial struggle being largely neglected.[5] Yet, Kenyan nationalism "lived more stories than the one we thought we knew",[6] not only within the Kenyan African population but also including Kenyan South Asians. Their voices and experiences in the Kenyan nationalist movement, their constituent part in this movement as a diaspora community and their transnational links not only shaped the Kenyan

4 After World War II, roughly 5 million Africans, 97,000 Asians and 29,000 Europeans lived in Kenya. In 1951, 11 elected white settlers, 7 Asian representatives and 4 hand-picked, nominated Africans were on the Legislative Council. Cooper calls the impact of events in South and South East Asia the "Asian roots of an African dilemma". Frederick Cooper: Reconstructing Empire in British and French Africa. In: *Past & Present Supplement* 6 (2011), pp. 196–210, here p. 199.

5 Bruce Berman / John Lonsdale: *Unhappy Valley. Conflict in Kenya and Africa*, 2 vols. Athens / Ohio: Ohio University Press 1992; David Anderson: *Histories of the Hanged. Britain's Dirty War in Kenya and the End of Empire.* London: Norton 2005; Caroline Elkins: *Britain's Gulag. The Brutal End of Empire in Kenya.* London: Pimlico 2005, and others.

6 John Lonsdale: KAU's Cultures: Imaginations of Community and Constructions of Leadership in Kenya after the Second World War. In: *Journal of African Cultural Studies* 13,1 (2000), pp. 107–124, here p. 108.

nationalist movement to a significant degree, but also contributed to the transimperial / transnational connections between East Africa and India at the time.

The 'classical' historiography on East African Asians has tended to concentrate on social, economic, and anthropological aspects rather than political involvement of South Asians in East Africa.[7] More recent work tends to focus on ethnographic elements and gender issues, for instance Herzig's study on South Asians in Kenya.[8] South Asians in Tanzania have been situated more clearly in the political context, particularly in Brennan's and Nagar's research.[9]

The role of South Asians in the anti-colonial struggle in East Africa does emerge in some recent biographies, for instance, those of Alibhai Mulla Jeevanjee, the founder of the East African Indian National Congress, and Makhan Singh, who formed the Labour Trade Union of Kenya in 1935 and the East African Trade Union Congress in 1949, by Zarina Patel. There is also a biography of Pio Gama Pinto, a leading South Asian Kenyan activist, in an account directed at children and young adults by Pheroze Nowrojee.[10] Moreover, several personal

7 Agehananda Bharati: *Asians in East Africa: Jayhind and Uhuru.* Chicago: Burnham 1972; Robert G. Gregory: *South Asians in East Africa: An Economic and Social History 1890–1980.* Boulder: Westview Press 1993; id.: *India and East Africa: A History of Race Relations within the British Empire, 1890–1939.* Oxford: Oxford University Press 1971; Jagjit S. Mangat: *A History of the Asians in East Africa c.1886 to 1945.* Oxford: Oxford University Press, 1969; Harold S. Morris: *The Indians in Uganda: Caste and Sect in a Plural Society.* London: Weidenfeld & Nicolson 1968; Cynthia Salvadori: *Through Open Doors: A View of Asian Cultures in Kenya.* Nairobi: East African Educational Publishers 1989.

8 Pascale Herzig: *South Asians in Kenya: Gender, Generation and Changing Identities in Diaspora.* Münster: Lit 2007; Gijsbert Oonk: *Asians in East Africa: Images, Histories & Portraits.* Arkel: SCA Producties 2004; Rasna Warah: *Tripple Heritage: A Journey to Self-Discovery.* Nairobi: Warah 1998.

9 James R. Brennan: South Asian Nationalism in an East African Context: The Case of Tanganyika, 1914–1956. In: *Comparative Studies of South Asia, Africa and the Middle East* 29,2 (1999), pp. 24–39; Richa Nagar: Saboteurs? Or Saviours? The Position of Tanzanian Asians. In: *Samar* 13 (2001), pp. 14–19; id.: The Politics of Gendered Boundaries: South Asian Communities in Tanzania. In: *Ghadar* 1,1 (1997), pp. 6–7, 15.

10 Makhan Singh: *Kenya's Trade Unions, 1952–1956.* Nairobi: Uzima Press 1980; id.: *A History of Kenya's Trade Union Movement to 1952.* Nairobi: East African Pub. House 1969; Zarina Patel: *Challenge to Colonialism. The Struggle of Alibhai Mulla Jeevanjee for Equal Rights in Kenya.* Nairobi: Publishers Distribution Services 1997; ead.: *Unquiet. The Life and Times of Makhan Singh.* Nairobi: Zand Graphics 2006; Pheroze Nowrojee: *Pio Gama Pinto. Patriot for Social Justice.* Nairobi: Sasa Sema Publications 2007. The Labour Trade Union of Kenya was renamed the Labour Trade Union of East Africa in 1937.

memoirs highlight some less well-known aspects of the relationship between Asians and Africans during that period.[11] Of course, South Asians and Africans did not constitute monolithic blocs, but had among them—in political terms—conservatives, moderates, and so-called radicals, as well as all shades of opinions between them. This paper focuses on the 'radicals', who have received less attention than the 'conservatives' and 'moderates' so far.

Besides academic scholarship, the participation of South Asians in the nationalist movement in East Africa has generated a wide public debate in the 2000s. A recent manifestation, the Murumbi Peace Memorial Garden, is situated in the vicinity of the City Park of Nairobi, encompassing the graves of Joseph Murumbi and his wife Sheila. Sculptures of Kenyan and Ugandan artists alternate with walls that are used as exhibition boards that narrate Murumbi's life, his involvement in the nationalist movement and in Kenyan politics[12]—as well as his indefatigable endeavours in collecting Kenyan artefacts. The choice of African Heritage Day for the opening of this peace memorial suggests that at least some influential individuals are trying to (re-)claim Murumbi as one of Kenya's 'freedom fighters'.[13]

One of the exhibition walls is dedicated to the memory of another Kenyan Goan in the nationalist movement, Pio Gama Pinto, who is buried nearby and who was the only non-African Kenyan honoured in a series of stamps entitled 'Heroes of Kenya' released in 2008. Moreover, he was posthumously awarded the Milele (Lifetime Achievement) Award in February 2007.[14] A few weeks before that, in January 2007, a poster featuring his portrait was displayed over the stage at the International Social Forum. It bore the caption *shujaa wa Kenya* ('hero

11 For instance, Mervyn Maciel: *Bwana Karani*. London: Merlin 1985; Manubhai Madhvani: *Tide of Fortune: A Family Tale*. Noida: Random House 2009.

12 Foreign Minister of Kenya 1964 to 1966, Vice-President May 1966 to December 1966.

13 African Colours, Keeping Murumbi's Legacy Alive, April 9, 2008; Invitation by the Murumbi Trust for the opening or the Murumbi Peace Memorial Garden on March 29, 2009.

14 Release date of stamps: 17 October 2008. For the Milele Award, see Kenya National Commission on Human Rights (KNCHR): *4th Annual Human Rights and Democracy Awards Report* (March 26, 2007), p. 5; Nowrojee: *Pio Gama Pinto*, p. 63. In addition, Pinto was awarded the Jaramogi Oginga Odinga Foundation's first Uhuru Award in July 2005. See ibid., p. 6.

of Kenya'). This epithet comes as a surprise when we consider the silencing of Pinto's work in the literature on nationalism in Kenya.[15] These examples together suggest a significant shift in how Kenyan nationalism and the individuals striving and fighting for it are publicly displayed.

So far, the story of the Murumbi Peace Memorial Garden or the labelling of Pio Gama Pinto as 'hero of Kenya' do not seem to be different from other processes of national commemoration.[16] However, the Murumbi Peace Memorial Garden as a site of such public national commemoration occurred only after a prolonged public debate of what to do with the area around the graveyard. Following a press release announcing the building of a huge shopping mall by a developer on the very spot of Murumbi's grave—land he had bought from the government—there was a huge outcry in the media. This eventually led to a reconsideration of the building plans and the establishment of a public memorial garden.[17]

Why are these episodes significant? In the 2000s, in contrast to the four decades following independence, the standard narrative of who was part of and contributed to the nationalist movement has been modified to include the roles of individuals of South Asian descent. In analysing this shift towards acknowledgement and inclusion of the flourishing transimperial / transnational connections across the Indian Ocean, I draw on interviews with Goans that I conducted in East Africa and elsewhere in addition to published and archival sources.[18]

15 South Asians frequently do not appear in the analysis of socio-political phenomena in Kenya at all: see e. g., Charles Hornsby: Social Structure of the National Assembly in Kenya, 1963–1983. In: *The Journal of Modern African Studies* 27,2 (1989), pp. 275–296. See also KNCHR: *4th Annual Human Rights and Democracy Awards Report*, p. 5.

16 Pierre Nora: *Les lieux de mémoire*. Paris: Gallimard 1984; Aleida Assmann: *Erinnerungsräume: Formen und Wandlungen des kulturellen Gedächtnisses*. 5th ed. München: C. H. Beck 2010.

17 Muriithi Mutiga: Nationalist Murumbi's Grave Vandalised. In: *Sunday Standard*, May 1, 2005; Patrick Mathangi: Murumbi Burial Site Grabbed, Claims Trust. In: *The Standard*, November 9, 2006; Allocation of Murumbi Grave a National Shame. In: *The Standard*, November 11, 2006; Peter Thatiah: Collector par Excellence. In: *The Standard*, November 18, 2006; Tony Kago: Alarm over Planto Grab the Grave of Former VP. In: *The Nation*, January 18, 2007; Murumbi's Grave Now Protected. In: *The Nation*, January 20, 2007; Gallery Befitting Murumbi. In: *The East African*, February 6, 2007.

18 Interviews were conducted between 2004 and 2009 in Kenya, Uganda, Tanzania, Zanzibar, Goa, Canada, Germany, and the UK with South Asians, mainly Goans, who

Transnational movement of ideas and people

There was a lively exchange of ideas between India and East Africa between the 1920s and the 1960s. These ideas, of course, had increasing weight as they were transformed from lofty utopias to an achievable reality once it became clear that India would gain its own independence from colonial rule. Several of the key contributors to these transnational exchanges moved between the Indian sub-continent, the different parts of East Africa, and the metropolis. Most were South Asians who received their education in India and the UK, and others were African Kenyans who studied in India or the UK. Here, I concentrate on the South Asians, in particular on Joseph Murumbi and Pio Gama Pinto, and their intellectual contributions to the nationalist movement in Kenya.

During the late 1940s and early 1950s, there was a significant shift in how political ideas, including the classical demands for *liberté*, *egalité* and *fraternité*, were expressed in the limited colonial and censored public sphere in Kenya. As Joseph Murumbi put it:

> The atmosphere was different, people were more conscious of their rights and more conscious of the need for change. They wanted a legislature in their hands, the elimination of the colour bar, more land, and more money spent on education and agriculture. When I left [Kenya for Somalia in1941], these demands weren't really being expressed.[19]

This transformation of the political atmosphere in the late 1940s and early 1950s received major inputs through the transimperial / transnational connections of local South Asians. Major global events such as the 'jewel in the crown' becoming independent cannot be underestimated, as well as the weakened economic position of the UK after the Second World War.[20] Furthermore, the example of Mahatma Gandhi, the non-cooperation and civil disobedience movements as well as his strategy of *satyagraha* and vision of *swaraj* had already had an impact

lived or still live in East Africa. To protect individuals' privacy, all interviewees are referred to with a pseudonym. The location mentioned is the location in which the interviewees were interviewed; this does not necessarily reflect where they live.

19 Anne Thurston: Emergence of a Nationalist: An Interview with Joseph Murumbi, part II. In: *Kenya Past and Present* 11 (1979), pp. 23–28, here p. 25. Murumbi returned to Kenya from Somalia in early 1952. The Human Rights Declaration was passed in 1948.

20 John Darwin: *The Empire Project. The Rise and Fall of the British World-System, 1830–1970*. Cambridge: Cambridge University Press 2009, pp. 514–565.

of how people thought and acted, not only in South, but also in East Africa.

The life and work of Joseph Murumbi and Pio Gama Pinto—only two examples of a wider range of South Asians who embodied transimperial / transnational connections and who engaged and involved themselves in Kenyan politics in the 1950s and 1960s—illustrate the lively to and fro between East Africa and India from the 1930s to the 1960s, and the use they made of existing personal and political contacts to exchange ideas across political boundaries, be they imperial or, later, national. Of course, as mentioned above, neither Kenyan Africans nor Kenyan Indians constituted a monolithic bloc with homogenous opinions: conservatives, moderates, and radicals could be found in any of these groups. For Kenyan Indians, an inhibiting factor in accessing their opinions now is that many of them worked in government departments and thus would risk losing their jobs if they criticized the government publicly or took up prominent political positions. For the historian, it is difficult to assess the different voices equally, because the sources available mostly relate to the more extreme positions of the spectrum, leaving a gap in the material relating to the section of the population that appeared to be aloof from politics.

Joseph Anthony Zuzarte Murumbi was born in Kenya in 1911 and died in 1990. His father initially worked as a clerk in the colonial administration when he arrived in East Africa in 1897, but later opened shops in several small towns in the Rift Valley, where he met Joseph's mother, the daughter of Murumbi, the Laibon (ritual and medical leader) of the Uasin Gishu Maasai.[21] When Joseph was about six years old, his father took him to India for his education. He learned to speak Hindi fluently, and after completing his studies, worked in famine relief in South India. According to him, this is when his political conscience awakened—by observing life in villages and how both social practice and the colonial authorities constructed barriers between individuals of different backgrounds. After returning to Kenya in the early 1930s, he lived in a number of towns and repeatedly experienced being labelled by one or other colonial administrative unit in racial categories as an 'Asian' or

21 Anne Thurston: Formation of a Character: An Interview with Joseph Murumbi, part I. In: *Kenya Past and Present* 11 (1979), pp. 16–21, here p. 18.

as an 'African' depending on the context.[22] This compartmentalization of colonial Kenyan society remained an issue throughout the struggle for independence and afterwards. As in India, in East Africa the colonial government labelled individuals in controversial ways, overriding the objections of individuals like Murumbi to their multiple identities being forced into a single category.

In the early 1950s, Murumbi met Pio Gama Pinto for the first time at the meetings of the 'Patel Brotherhood'. Pinto invited Murumbi to join the Kenya Study Circle, comprised of Kenya African Union (KAU) members including Bildad Kaggia, Achieng Oneko, and Paul Ngei. At one of those meetings, Murumbi also met Jomo Kenyatta, and decided to join the Kenya African Union. He became the KAU's Acting Secretary and was sent out "to tell the world what was happening in Kenya".[23] Murumbi's 'mission' took him to India first, on to the UK, and should then have led back to Kenya via the US. However, he was denied a visa to the US, and was not able to return to Kenya, because the KAU had been banned by the colonial government there. During the nine years (1953–1962) Murumbi spent in the UK, he established extensive transnational networks in support of the anti-colonial struggle. Murumbi describes this period metaphorically: "I always think that we were like little termites burrowing into the structure of the colonial edifice."[24]

Not only were these networks useful when the Lancaster House conferences took place in London in the early 1960s, they also provided global publicity for the Kenyan nationalist perspective. After independence, Murumbi became Kenya's first foreign minister and later on, its second vice-president, from May 1965 to August 1966. Following his

22 Thurston: Interview with Murumbi, part II, p. 24. In one instance, his case was brought before the Attorney General who "ruled that [he] had the option either to stick to my Asian nationality and give up any rights as an African, or vice versa".

23 Thurston: Interview with Murumbi, part II, p. 26. The 'Patel Brotherhood' was founded by Indians of Gujarati origin in Nairobi in 1920.

24 Thurston: Interview with Murumbi, part II, p. 28. These networks included members of the Congress of Peoples Against Imperialism (CPAI; chairman was Fenner Brockway) and the Movement for Colonial Freedom. Both Jomo Kenyatta and Kwame Nkrumah were frequent visitors to Swaraj House in London, where the CPAI met. The National Archives, London, Public Record Office (hereafter PRO), KV 2/1789.

resignation from politics, he concentrated on collecting African art and literature about Africa.[25]

However, the nationalist movement could not function on ideology alone: hand in hand went the practice of nationalism, which required material means to follow through the intellectual, ideological, and political demands the Kenyans made against the British colonial government. The South Asian diaspora in Kenya contributed not only to the ideological thrust of the nationalist movement, but invested major sums of money in the anti-colonial struggle. In East Africa, South Asian businessmen financed several journals and presses which were essential for spreading the ideas of a free Kenya. In the interviews I conducted, individuals frequently mentioned the financial support and private initiatives they provided to various undertakings. Apart from local South Asians, the Indian government supported Indian individuals in East Africa, and African individuals coming to India for their studies.[26] For instance, the Indian government paid Murumbi a weekly allowance in the 1950s in order to "go around England speaking about Kenya".[27]

India's first prime minister, Jawaharlal Nehru, strongly advocated that every country have its independent and self-elected government, including the East African countries. Nehru seems to have been rather well-informed about the political situation in East Africa—which might be partly due to the exchange between the East African Indian National Congress and the Indian National Congress (see below). On a global scale, Nehru initiated a policy of non-alignment and *pancasila*.[28] These were formally put in place at the Bandung conference in 1955.

25 Thurston: Interview with Murumbi, part III, p. 31

26 PRO, KV 2/1788; Interview with Samuel, Kenya, August 2006. Nehru introduced scholarships for Africans to study in India. One of the first scholarship awardees from Kenya was Omamo, Odinga's cousin.

27 Thurston: Interview with Murumbi, part II, p. 26.

28 Pancasila: five principles of peaceful co-existence: mutual respect for each other's territorial integrity; mutual non-aggression; mutual non-interference in internal affairs; equality and mutual benefit; peaceful co-existence (Nehru). Pancasila formed the basis ofthe strategy of non-alignment. For a comparative analysis of the non-alignment movement see Akhil Gupta: The Song of the Non-aligned World: Transnational Identities and the Reinscription of Space in Late Capitalism. In: *Cultural Anthropology* 7,1 (1997), pp. 63–79.

Another South Asian who influenced the nationalist movement in Kenya very strongly was Pio Gama Pinto, who was born in Nairobi in 1927. Pinto's father held a clerical job in the colonial administration, like so many Goans living in East Africa. Pinto was sent to India for his education, where he stayed from 1938 to 1949. Despite his youth he was active in political circles: Pinto was one of the founder members of the Goa National Congress, the aim of which was to free Goa from Portuguese rule.[29] After returning to Nairobi in 1949, Pinto involved himself in the East African Indian National Congress (EAINC), the Kenya Freedom Party (KFP) and the Kenya African Union (KAU), working mostly behind the scenes—drafting programmes, memoranda, and manifestos for all these political groups, including the Kenya African National Union (KANU) later on. He also drafted the reports for the Kenya Land Commission, for fund-raising for the nationalist movement, and to support ex-detainees. His fervent wish was to improve the lives of Africans and Indians in Kenya and to gain independence from colonial rule.[30] Pinto cooperated with the representatives of the Trade Union, particularly Makhan Singh and Fred Kubai; he also enjoyed a friendly relationship with Apa Pant, India's High Commissioner in Nairobi. His activities were interrupted between 1954 and 1959, when Pinto was detained under the emergency regulations.[3131] His transnational links across empires and experiences that he collected in Goa and in India in the 1940s would benefit the nationalist movement in Kenya in the 1950s and 1960s.

As soon as Kenya achieved independence in 1963, Pinto and his wife took up Kenyan citizenship. Pio Gama Pinto was one of 12 specially elected members of the House of Representatives and won a seat at the East African Central Legislative Assembly, and in this role delivered several critical speeches.[32]

29 Oginga Odinga et al.: *Independent Kenya's First Martyr. Socialist and Freedom Fighter – Pio Gama Pinto.* Nairobi: Pan Africa Press 1966, pp. 5, 29.

30 Ibid., pp. 35, 28; Interview with Adela, Canada, July 2007; Interview with Alberto, UK, May 2005.

31 Pinto was detained at Manda Island from 1954 to 1958. He was the only non-African there and was kept in isolation. PRO, CO822/1233. From 1958 to 1959, Pinto was kept in restriction at Kabarnet.

32 Letter from Pio Gama Pinto to all KANU members of the House of Representatives, 29 May 1963, in Kenya National Archives (hereafter, KNA), MAC/KEN/71/1. See also Kenneth Good: Kenyatta and the Organization of KANU. In: *Canadian Journal of African Studies* 2,2 (1968), pp. 115–136, here p. 119.

This is reflected in the statements of a contemporary of Pinto's when asked about Pinto's vision and fears for an independent Kenya:

> Pinto was very worried [. . .] used to say this many times—that [if] at the end of the exercise—by getting freedom—we get a lot of black Lord Delamares ruling this country rather than white Lord Delamares—then we would not have achieved very much—and I feel that he was right.[33]

In early 1965, Pinto was advised by friends to leave Nairobi for an extended period because they perceived him to be in danger. On 24 February 1965, the day after he returned from his self-imposed stay in Mombasa, Pio Gama Pinto was assassinated.[34] Pinto's assassination was the first political murder in independent Kenya and sparked lasting and controversial debates as to who planned it. The man who was convicted for Pinto's murder seems to have been a 'fall guy'. To this day, speculations about the background of the assassination run high.[35]

According to many contemporaries, Pinto was highly appreciated by a wide circle of friends and colleagues among the Kenyan African and Kenyan Indian politically engaged. As Muinga Chokwe put it: "People in Mombasa used to say that but for his colour they would have thought him to be my brother. Such was Pio's nationalist fervour."[36] Not only did Pinto stand for an independent Kenya and an egalitarian society, which was his vision for the nation-building project, he was also perceived as the main strategist and intellectual in left-of-centre

33 Interview with Alberto, UK, May 2005.

34 This came as a great shock to politicians, and particularly the South Asian community in Kenya: Interview with Alberto, UK, May 2005. Pinto's papers were all burnt in his office by a circle of friends in order to protect his friends and colleagues from any anticipated consequences. Many authors now quote 25 February 1965 as the date of Pinto's assassination, but this is not correct.

35 Kamau Ngotho: Kenya: The Pinto Murder Mystery Resolved. In: *The Daily Nation*, June 18, 2000; Kenya: How Pinto Murder Was Plotted … And Kisilu Framed. In: *The Daily Nation*, June 19, 2000. Kamau Ngotho received the Developing World Award by the EU for this article. François Lefèbvre: The Lorenzo Natali Prize: Human Rights and Development, the Real Winners. In: *The Courier ACP-EU* 190 (January–February 2002), pp. 1–90, here p. 30. A series of further political assassinations followed; assassinated were among others Tom Mboya, J. M. Kariuki, and Robert Ouko.

36 Odinga: *Independent Kenya's First Martyr*, p. 36. However, Pinto was perceived and received with sceptical distance in the wider South Asian community. This was particularly true for Goans, many of whom felt bound to keep the loyalty to the colonial government as they had to vow this when entering the administrative service.

political circles.[37] He drew on his transimperial / transnational links to a great extent; they provided intellectual as well as material support for the anti-colonial liberation struggle.

Apart from Joseph Murumbi and Pio Gama Pinto, other South Asian personalities involved in anti-colonial circles at the time included Apa Pant, the first Indian High Commissioner in Nairobi, who served there from 15 August 1948 to late February 1954 and was characterized as "Nehru's personal emissary in East Africa".[38] Apa Pant had to leave Kenya when he was declared *persona non grata* and sent home by the British government. Pant actively inspired Kenyan Indians at every opportunity "to fight against racialism, sectarianism, and communalism, and to give shelter to Africans during Mau Mau"; and "encouraged them to take up citizenship in their country of adoption".[39] The combination of different levels of transimperial / transnational connections seemed to have worked well: Pant represented the diplomatic level, Pinto and Murumbi the local input of the South Asian diaspora to the Kenyan nationalist movement.

Several more initiatives connected Kenyan Africans and Kenyan Indians in the struggle for independence. One of them was the organization of the defence for the so-called Kapenguria Six.[40] Three of the seven lawyers representing Jomo Kenyatta and the other accused were Kenyan Indians; one of the seven came directly from India for the trial.[41] Particularly the Kenyan Indians risked their reputation if not their lives in order to collect evidence and find witnesses throughout Kenya in favour of their clients. Despite this prolific team of lawyers, the case was lost and all six were convicted and sentenced to imprisonment for

37 See Robert G. Gregory: *Quest for Equality. Asian Politics in East Africa 1900–1967*. Hyderabad: Orient Blackswan 1993.

38 Dana April Seidenberg: *Uhuru and the Kenya Asians: The Role of a Minority Community in Kenya Politics, 1939–1969*. New Delhi: Vikas 1983, p. 79. Pant's first day in his office in Nairobi was the first anniversary of Indian Independence.

39 Ibid., p. 119. See also PRO, KV 2/1788.

40 Jomo Kenyatta (c.1894–1978), Bildad Kaggia (c.1921–2005), Kung'u Karumba (unknown; gone missing in 1975), Fred Kubai (1917–1988), Achieng Oneko (1920–2007), and Paul Ngei (1923–2004).

41 The team of defenders consisted of D. N. Pritt (1887–1972), Chaman Lall (who had been requested by Nehru to go to East Africa for this purpose), A. R. Kapila (1926–2003), F. R. S. de Souza (alive 1929), and Jaswant Singh, one lawyer from Jamaica and H. O. Davies (1905–1989) from Nigeria.

up to ten years. Nevertheless, these and other Indian lawyers continued to defend hundreds of Mau Mau suspects in courts throughout Kenya in the 1950s. The example of the practice of nationalism—not just ideological and financial, but also legal—emphasizes the transimperial / transnational links across Africa and across the Indian Ocean, including the active support of the Indian government.

A second initiative was the 'Free Kenyatta Campaign' which was launched and seen through by Ambu Patel (1919–1977) in Kenya in October 1952. Patel described himself as a 'Gandhian', and wrote about his campaign:

> [...] to the end of 1961, I have lived, worked, and breathed every moment of the day for Baba's [Kenyatta'a] release and Kenya's freedom. I have adopted Jomo Kenyatta as my Baba (father) and all other Africans as my brothers and sisters.[42]

Ambu Patel and his wife Lila (1927–1979) organized many demonstrations demanding Kenyatta's release, for example one at Kiambu, which was well attended by Africans and Indians. The couple also wrote and circulated pamphlets to politicians and well-known public figures all over the world, invoking a wide transnational community in order to get 'moral support' for their campaign. However, the British colonial authorities in Kenya censored the correspondence and held back international replies. Finally, when a few of them were passed on to the Patels, some of the supportive messages were incorporated in posters under the heading "World's Voice of Release of Jomo Kenyatta" that were displayed for eleven weeks throughout the country. With this campaign, Patel created a global public arena for the demands of the Kenyan nationalists, and especially, the release of Jomo Kenyatta. Besides raising the awareness of the need to release Kenyatta, Patel sent him food and clothes "at great personal risk" and also looked after Kenyatta's daughter Margaret during the years of Kenyatta's detention.[43] Kenyatta was finally released in October 1961. Thus, Patel combined ideological support for the nationalist movement with very tangible democratic and material assistance.

42 Ambu Patel: *Struggle for Release Jomo & His Colleagues*. Nairobi: New Kenya Publishers 1963, p. 32.

43 Seidenberg: *Uhuru*, p. 118.

Although tensions and different perceptions existed among African Kenyans and Indian Kenyans with respect to how to follow through the liberation movement, both groups emphasized the importance of cooperation. The examples of members of the South Asian diaspora in East Africa who actively engaged in the anti-colonial movement in Kenya discussed here—Murumbi, Pinto, Pant, and Patel—are complemented by a statement of Jomo Kenyatta, to name but one example. He suggested in a newspaper article that Indians and Africans should not be misled by the divisive policies employed towards the colonized by the British government, but should work together on the basis of their common interest:

> Indians and the Africans have politically many things in common and a common platform can be built where the members of the two communities can work together on *completely equal footing*.[44]

Kenyatta's emphasis on the equality between Africans and Indians in the anti-colonial struggle and his vision of a future together confirm the impression that the input of the South Asian diaspora to the nationalist movement, be it intellectual, political or financial, was seen by at least a significant number of the most prominent African leaders as essential to its progress and subsequently, success. It correlates with Harry Thuku's statements on cross-fertilization between the Indian and Kenyan nationalist movements, the ideological and symbolic importance of Gandhi's campaigns, and thus, the transimperial exchanges of ideas. This is reflected in Kenyatta's attendance at several meetings organized by the India Youth League in Nairobi and at the celebrations of Indian Independence Day in Nairobi on 15 August 1947.[45]

Thus, personalities such as Jeevanjee and Thuku, Murumbi and Pinto sowed the seed for establishing a public arena for demands of independence, democracy, and equality. Their individual involvement benefited

44 *Colonial Times*, October 12, 1946. PRO, KV 2/1788.

45 PRO, KV 2/1788. "In celebration of Indian Independence Day a mass meeting of Indians was held in Nairobi under the auspices of Indian public bodies and was also attended by many Africans. After speeches by Indian leaders Jomo Kenyatta praised Indians for what they had achieved and stated that he would ensure that his community followed their example with regard to African independence. Africans, he said, would require the assistance of the Indian who would receive the support of the African too in the future."

various organizations, mostly political parties, but the press and trade unions as well, which offered a platform to voice the demands. During the Second World War, political activities were at a low, but in the late 1940s and 1950s, democratic demands were more strongly articulated. However, these were squashed by censorship. In the 1950s, the demands for a fairer land distribution and with it, Mau Mau and the Emergency were the dominant issues in Kenya. Some political activities were linked to Mau Mau, but did not necessarily support the insurgents—but this is a different topic, not discussed here.[46] In the following, I will concentrate on how institutions displayed nationalist positions.

Manifestation of ideas in institutions

How were transimperial / transnational connections and the ideas of *liberté*, *egalité* and *fraternité* institutionalized in Kenya? During the years leading to Kenyan independence, several organizations and institutions, particularly political parties, the press and trade unions, were established by Kenyan Indians and Kenyan Africans. In these, ideas of equality and universal rights for everyone—and for every nation-state—were manifested as well.

Political parties

The first political party to be founded in Kenya was the East African Indian National Congress (EAINC) in 1914 by Alibhai Mullah Jeevanjee. It was shaped along the lines of the Indian National Congress (INC).[47] Initially, the EAINC was concerned with matters relating to the South Asian population of East Africa. The main demands of the EAINC were to defend and advance the rights of Indians in East Africa with respect to property, business, and representation on the Legislative Council, in particular the right to franchise; the removal of restrictions on land sales; greater representation of Indians on legislative and municipal councils; the right to trial by jury and the condemnation of European settlers' demand for the restriction of Indian

46 A vast body of literature analyses Mau Mau and the Emergency. For the latest study on this topic, see Daniel Branch: *Defeating Mau Mau, Creating Kenya: Counterinsurgency, Civil War, and Decolonization* (= African Studies, vol. 111). Cambridge / New York: Cambridge University Press 2009.

47 The Indian National Congress was founded in 1885.

immigration into the protectorate.[48] In the 1930s, the focus of the EAINC shifted to broader issues of self-determination in Kenya and to creating an independent Kenya for all. Members of the EAINC and the INC moved constantly between East Africa and South Asia; they took part in the debates of the INC, in Mahatma Gandhi's anti-colonial demonstrations and *satyagraha* campaigns. Indians came to preside over the meetings of the EAINC, as did, for instance, Sarojini Naidu in 1924 and in 1930. Her presidential addresses reflected on the situation of the anti-colonial movement in India and gave inspiration to her East African counterparts.[49] In addition, members of the EAINC had experienced Gandhi's political strategies of civil disobedience and non-cooperation, which culminated in the *Quit India* movement of 1942. These transimperial interactions illustrated that the ideas of demanding self-government, democracy, and equality were not merely lofty utopias, but would eventually lead to the goal of breaking free from colonial rule and establishing an independent nation-state.

In 1952, the EAINC was renamed the Kenya Indian Congress (KIC) as it restricted its operations to Kenya. More significantly, the EAINC / KIC and the Kenya African Union (KAU) held common meetings in which they underscored their goal to obtain a greater say in government affairs and their struggle for a common cause: Kenyan independence and a democratic society. This was as an attempt to integrate transnational links into the nationalist movement in Kenya, which arguably could not have otherwise been sustained.

The short-lived East African Association[50], active in 1921 and 1922, was the first multi-ethnic association with an ideology of equality for all. Its founder Harry Thuku (1895–1970) organized a succession of events under its umbrella in various districts in Kenya during those two years. In his speeches, Thuku suggested that Africans should throw away their *kipandes*[51], pay a lower hut tax, and refuse to do unpaid labour including

48 Presidential Address by T. M. Jivanjee, President First Session, 1914, KNA MAC/EAI/28/1.

49 Presidential Address by Shrimati Sarojini Naidu, President Fifth Session, 1924, KNA, MAC/EAI/28/1. Naidu also presided over the Ninth Session of the EAINC held in Nairobi in 1930.

50 Formerly the Young Kikuyu Association, founded in 1920.

51 The *kipande* was a box containing identity papers including the name, fingerprints, and employment history of the bearer. It had to be carried around the neck and was

roadwork. Furthermore, Thuku invoked the example of Gandhi and averred that he wanted to be like Gandhi. However, his activities were limited to a short time period: in spring 1922, Thuku was imprisoned for "threatening peace and good order" under the Removal of Natives Ordinance. Following a peaceful prayer demonstration of 8,000 individuals for his release, at which the police and white settlers fired on the demonstrators, he was deported without trial to Northern Kenya.[52] Keith Kyle argues that apart from being the end of the East African Association and the peaceful movement to request constitutional changes, the harsh reaction by the colonial regime "[broke] the potential alliance between Indians and Africans".[53]

In contrast to Kyle, I would argue that there were at least partially successful attempts in the 1940s and 1950s to establish just such an alliance between Asians and Africans to gain *swaraj* (self-rule) in Kenya through demands for peaceful constitutional changes, for instance with the Kenya African Union (KAU), founded in 1944. This political organization followed a pan-ethnic programme; it continuously petitioned for self-government and collaborated with the EAINC. Kenyatta was president of the KAU from 1947. According to John Lonsdale, the young leaders of the KAU in the early 1950s were convinced of being able to create a multi-racial and multicultural society—not something the settlers wished for, as they preferred to defend privileges which were closely linked to the policy of racial segregation.[54]

The cooperation between the KAU and the EAINC underscored their goal to obtain a greater say in government affairs, and to establish *swaraj* for Kenya eventually. After joint meetings of the KAU and the EAINC were banned by the colonial government in April 1952, the hitherto promising anti-colonial joint venture between Kenyan Africans and Kenyan Indians came to an end. At their last meeting the

made compulsory for every African Kenyan over the age of 15 through the Native Registration Amendment Ordinance 1920.

52 East Africa. Papers Relating to Native Disturbances in Kenya (March, 1922). Parliamentary Papers 1922. Cmd 1691 (London, 1922), *passim*. Apparently the Chief Native Commissioner was particularly concerned about the religious aspect of Thuku's speeches and therefore initiated and implemented Thuku's deportation. Thuku was released only in 1931.

53 Keith Kyle: Gandhi, Harry Thuku, and Early Kenya Nationalism. In: *Transition* 4,27 (1966), pp. 16–22, here p. 22.

54 Lonsdale: KAU's Cultures, p. 116.

EAINC and KAU members stated that the ban on their joint meetings was "liable to be construed as expressing the sympathy of the Kenya Government with the racial policies of the Union of South Africa".[55] Shortly afterwards, in 1953, the KAU was banned as a party, because the British perceived the KAU to be a cover organization for Mau Mau. They convicted the KAU's leading officials, among them Jomo Kenyatta, accusing them of instigating and managing Mau Mau.

The KAU has been evaluated differently by scholars: the general historiography depicts the KAU as a failure in terms of high politics, organization, class representation, and leadership. Lonsdale takes a slightly different view: he suggests that the KAU could not succeed in establishing African majority rule as the British did not listen; the land issue was not solved through negotiations; and the internal make-up of the KAU was uneven. It therefore lost the support of young and poor Kenyans because it did not achieve results with respect to African representation and the land issue.[56] Here, I would like to add another twist: the KAU might not have been successful in achieving tangible results such as solving the land issue, but it created a common platform for Kenyan African and Kenyan Indian politicians, it created an awareness of the need for change towards *swaraj* and *uhuru*, it created networks on which future negotiations for independence were taken forward, and it provided the basis for the nucleus of individuals who were to found the KANU in 1960.

During the Mau Mau and emergency years (1952–1960), political activities in Kenya were restricted in some years, partly due to the imprisonment or detention of personalities such as Pio Gama Pinto and Jomo Kenyatta in the mid-1950s, or the exile of Joseph Murumbi, and more generally, the very tight control and censorship that was exercised by the British colonial government on any activity undertaken in Kenya. Although the Lennox-Boyd Constitution of 1958 increased the numbers of individuals on the Legislative Council significantly, and for the first time included the election of African members, major demands towards self-government were not met. Thus, the struggle went on. Besides the Kenyan African National Union (KANU) and the Kenyan

55 Minutes of the Joint Meeting between the EAINC and the KAU in Nairobi, April 10, 1952, KNA, MAC/KEN/36/2.

56 Lonsdale: KAU's Cultures, pp. 109–110.

African Democratic Union (KADU), another smaller party group was founded in 1960—the Kenya Freedom Party (KFP), the last example that I discuss here. The KFP also reflects a close cooperation between Kenyan Africans and Kenyan Indians. The so-called radicals founded the KFP after having issued the Statement of the Thirty-One, which demanded the speedy formation of a democratic government in Kenya, modelled after the British example.[57] Accordingly, the specific aims of the KFP, which unlike the KADU or the KANU was open to everyone, irrespective of race, class, gender, caste, or ethnicity, were:

1. To work for the attainment of independence for Kenya with a constitution of the type that obtains in the United Kingdom, based on the principles of democracy with an Executive responsible to a legislature elected on a common roll based on universal franchise;
2. To work for the enactment of legislation guaranteeing to individuals human rights as contained in the Declaration of Human Rights approved by the United Nations Organization;
3. To work in particular for the establishment of equality of opportunity of the people of Kenya to enable them to make the fullest contribution to the development of the country.[58]

Apart from territorial issues (the demand for a unified Kenya), the KFP advocated equal opportunities for everyone to join the civil service irrespective of race or gender and the abolition of hierarchical pay-scales. They passed resolutions about the economic and social future of Kenya, which should lead to the "fair distribution of wealth so that the prosperity of the country is reflected in the man in the street".[59] Schools, hospitals, and other social facilities should be open to anyone; that is, the KFP demanded an integrated education and public health sector, which was a distinct difference from the policies of the British colonial government which were based on racial categories and hierarchies. The most prominent members of the KFP were Pio Gama Pinto, the party's assistant secretary, and Joseph Murumbi. One of the KFP members remarked: "We later on formed the Kenyan Freedom

57 Interview with Alberto, UK, May 2005.

58 KFP papers February 1960 as quoted in Seidenberg: *Uhuru*, p. 151.

59 KP Shah interview with Seidenberg 1977, as quoted in Seidenberg: *Uhuru*, p. 161.

Party and I think we did our little bit."[60] Furthermore, the KFP actively lobbied for Kenyatta's release throughout 1960–61. A delegation of six of its members visited Kenyatta in detention in Maralal. At this visit, Kenyatta "assured the [Kenyan Indian representatives of the KFP] that the future of the Asians in Kenya was secure".[61] This was in line with his statement emphasizing the significance of Kenyan Indians and Kenyan Africans working together for a common cause—a free Kenya.

The press

Apart from political parties, the East African Indian press played a major role in publicizing the ideas of nationalism and advocating *swaraj* and *uhuru.* The general emphasis in the majority of articles in newspapers such as the *Colonial Times* and the *Daily Chronicle* was to stress the need to join together their own efforts with those of African nationalists in the anti-colonial struggle and overcome the racial segregation policies that had been put in place by the British government. Pio Gama Pinto was one of the most prominent Indian journalists, and was editor of the *Daily Chronicle* besides his activities in various parties outlined above. He prepared numerous brochures and pamphlets demanding equal rights for everyone in Kenya and also worked as commercial manager of the Pan-African Press Ltd., which published *Sauti ya Mwafrika* and *Pan-Africa.* Oginga Odinga, the first Vice-president of Kenya, characterized Pinto as "the moving spirit in the establishment of Pan-Africa Press".[62]

An Indian printing press published and financed the African nationalist newspaper *Mwiguithania* edited by Jomo Kenyatta.[63] The Government reacted by charging the responsible editors and journalists with writing 'seditious' articles. The penalties included fines and their imprisonment.

60 Interview with Samuel, Kenya, August 2006. The Kenya Freedom Party was founded in 1960 and dissolved in 1963. *The Kenya Gazette*, May 31, 1960 (62,28), p. 683; *The Kenya Gazette*, April 16, 1963 (65,17), p. 419.

61 *East African Standard*, June 12, 1961.

62 Oginga Odinga: *Not Yet Uhuru. The Autobiography of Oginga Odinga.* London: Heinemann 1967, p. 270. The newspapers or journals published at the Pan-African Press included a weekly in Dhluo *Nyanza Times*, a weekly in Swahili *Sauti ya Mwafrika*, and a bi-monthly in English *Pan Africa.*

63 Mwiguithania = 'The Reconciler'.

Also, the Government introduced a law that prohibited the printing of 'seditious' articles.[64]

The Trade Unions

The ideas of equality and universal rights for everyone were also manifested in the formation of trade unions in the early twentieth century. The unions were of particular importance to the nationalist movement, as they did not only unite workers in order to demand equal pay for equal work, but also encouraged workers to demand political rights such as the adult franchise. The union was a non-racial organization, open to all workers irrespective of race, religion, caste, creed, or ethnicity. It thus was one of the very few organizations in colonial Kenya which was not based on racial segregation. Trade Unions were active from the 1930s onwards; the Labour Trade Union of Kenya was founded in 1937. The two major figures in the trade union movement were Fred Kubai (1917–1988) and Makhan Singh (1913–1973), a Kenyan African Indian team. Apart from aiming at organizing workers on a class basis and at improving their economic conditions, the union's main thrust was to demand universal rights. For instance, at the 3rd annual conference of the Labour Trade Unions in East Africa in 1939, the delegates passed a resolution demanding "freedom of speech, freedom of meeting and procession, freedom of press, freedom of movement, freedom of organization."[65] According to Makhan Singh, the "real significance lay in the fact that they [the resolutions] had resulted from the joint deliberations of African and Asian workers for the first time in East Africa."[66]

An umbrella institution of trade unions was formed in 1949: the East African Trade Unions Congress which specifically aimed at achieving better working conditions for Africans and Asians.[67] They also orga-

64 The Printing Presses (Temporary Provision) Ordinance of 1952 came into full operation once the Emergency was declared in October 1952. Accordingly, 12 newspapers were suppressed, among them *Sauti ya Mwafrika*, as well as a further 29 leaflets and pamphlets in October 1952. In addition, the Emergency (Publications) Regulations of 1953 controlled the licensing of newspapers and journals. F. Carter: 'The Kenyan Government and the Press', 1906–1960, KNA, MAC/KEN/64/1.

65 Singh: *A History*. Incidentally, these demands were to be included in the Human Rights Declaration of 1948.

66 Ibid., p. 82.

67 Ibid., p. 202.

nized a boycott of the celebrations of the granting of a Royal Charter to Nairobi in 1950 to demonstrate how the rapid development of Nairobi impinged on the lives of common people. This developed into a general strike demanding self-government for the East African territories along the lines of point 5 of Wilson's 14 points programme of 1918; the abolition of the *kipande*-system; the abolition of paying wages according to race; and an increased minimum wage.[68] After this event, the trade union was banned by the government. Fred Kubai and Makhan Singh were imprisoned.[69] For his activities in establishing and running the union, some of my respondents called Makhan Singh "the greatest union leader that Kenya has produced".[70]

These examples of political parties, the press, and trade unions show how ideas of democracy and equality manifested themselves in a south-south transimperial / transnational exchange along the Indian Ocean rim—and that more collaboration and cooperation between Kenyan Africans and Kenyan Indians took place than is generally acknowledged in the literature. At the same time, it is apparent that the divisive policies of the colonial regime limited this collaboration and how it was perceived, not only at the time but even up to the present day. Nevertheless, the cooperation between Kenyan Indians and Kenyan Africans had a major effect on Kenyan politics and its path towards Independence. Although after Independence, South Asians were accused of being aloof from politics in the nationalist movement, in practice, and relative to the size of their population, they contributed disproportionately in terms of individuals, ideas, finances, and political networks to the attainment of *swaraj* for Kenya.

Independence and beginnings of the nation-building process

When it was certain that Kenya would obtain independence by December 1963, the Kenya Indian Congress dissolved itself in mid-July 1963

68 Address delivered by Woodrow Wilson to Congress on 8 January 1918. Its main points were free trade, open agreements, democracy, and self-determination. See Erez Manela: *The Wilsonian Moment. Self-Determination and the International Origins of Anticolonial Nationalism*. New York: Oxford University Press 2007.

69 See Singh: *A History*. It is noteworthy that Makhan Singh spent 11.5 years in detention—the longest of any politically detained person during the colonial period in Kenya.

70 Interview with Sebastian, UK, May 2005.

after the first general election in May 1963. The general perception among Kenya Indian Congress members was that there was no need any more to "continue to function politically as an Asian organization".[71] In November 1963, the Kenya Freedom Party adopted the same line in the presence of—among other KANU leaders—Tom Mboya and Mwai Kibaki, after KANU had opened its doors to everyone irrespective of race.[72] It is not entirely clear from the available sources how much coercion was involved in dissolving these two comparatively successful parties, but these moves denoted the end of a strong Indian public voice in Kenya and the distinct visibility of Kenyan Indians. In retrospect, we can see this shift as a sign of fading transnationalism and a move towards an inward-looking nationalism.

Kenyatta's central role as first president was underscored by the dissolution of KADU in November 1964, thus de facto creating a one-party state.[73] Although the so-called KANU backbenchers acted as a critical voice in parliament, they did not last long.[74] The Kenyan government followed a policy of 'Africanization' in the civil service and the economy, reproducing colonial categories of racial segregation—between 'natives' and 'non-natives'—with the difference that in the post-colonial situation, colonial hierarchies were turned upside down.[75] Although the policy of 'Africanization' played a role in many newly established African nation-states, including Ghana, Cameroon, Nigeria, Kenya, and

71 *East African Standard*, July 23, 1962, as quoted in Seidenberg: *Uhuru*, p. 163.

72 Letter from Pio Gama Pinto to all KANU members of the House of Representatives, May 29, 1963, in KNA, MAC/KEN/71/1.

73 Daniel Branchand / Nicholas Cheeseman: The Politics of Controlin Kenya: Understanding the Bureaucratic-Executive State. In: *Review of African Political Economy* 33,107 (2006), pp. 11–31; Good: Kenyatta and KAU, p. 132.

74 See The Changing Face of Kenya Politics. In: *Transition* 25 (1966), pp. 44–50, here pp. 46, 50.

75 See for instance, Mahmood Mamdani: Beyond Settler and Native as Political Identities: Overcoming the Political Legacy of Colonialism. In: *Comparative Studies in Society and History* 43,3 (2001), pp. 651–664; David Himbara: Failed Africanization of Commerce and Industry. In: *World Development* 22,3 (1994), pp. 469–482. Tanzania retained a more transnational outlook than Kenya, linked to both Nyerere's ideas of following Pan-Africanism and socialism as well as the absence of a large white settler community. See Ronald Aminzade: The Politics of Race and Nation: Citizenship and Africanization in Tanganyika. In: *Political Power and Social Theory* 14 (2000), pp. 53–90; Festo Mkenda: *Building National Unity in Sub-Sahara Africa: The Impact of State Policies on the Chagga Community in Northern Tanzania*. University of Oxford DPhil thesis. Oxford 2009.

Tanzania, it was not necessarily a major part of the nation-building process. It was applied in varying degrees in these countries, and the implications of this shift are not easily deciphered, because it was connected to questions of staying in power and distributing benefits to different groups within those deemed to be 'African'. In the Kenyan case, 'Africanization' of the civil service and the economic sector proved to be a relatively easy method of restricting Indians from taking up jobs in the administration and requesting trade licences. Such moves seemed to promise economic redistribution results quickly and at the same time provided the general public with a 'scapegoat'. Access to power and state resources might have been perceived as linked to internal divisions along ethnic lines, and thus, racially based nationalism was seen as a uniting factor in an increasingly divided and fragmented society.[76] The process of 'Africanization' represented a shift from emphasizing 'nationalism in Africa'—inclusive of any person partaking in the venture of building a new nation-state—to 'African nationalism'—restricted to Africans, and this is how several of my interviewees saw the early years of the newly established nation-state in Kenya.[77]

From the mid-1960s onwards, Kenyatta more or less aligned his political strategies with 'western' wishes: more conservative social and foreign politics, characterized by an entrenchment of exclusionary strategies. Whilst Africanization was mainly motivated by short term political considerations, the policy also legitimized the re-emergence of colonial categories, perspectives, and policies. Discussions on 'African socialism' which were part of the pre-independence debates of the 'freedom fighters' did not play much of a role after the mid-1960s—in contrast to Tanzania, where Nyerere emphasized the concept of 'African socialism' in the nation-building process. As one of my interviewees put it:

76 For Tanzania, see Brennan: South Asian Nationalism; for Zanzibar, see Jonathon Glassman: Sorting out the Tribes: The Creation of Racial Identities in Colonial Zanzibar's Newspaper Wars. In: *The Journal of African History* 41,3 (2000), pp. 395–428; for Uganda, see Frank Schubert: Die kolonialen Wurzeln nachkolonialer Gewalt: Die Entstehung von Martial Tribes und politischer Ethnizität in Uganda. In: *Archiv für Sozialgeschichte* 48 (2008), pp. 275–294.

77 See Benedict Anderson: *Imagined Communities: Reflections on the Origin and Spread of Nationalism*. London: Verso 2006. Interview with Antonio, UK, May 2005.

> Kenyatta in fact turned completely. Once he was a strong socialist— [a] very strong socialist—in fact he was in a communist party in Russia. [But then] he became completely a private enterprise—I think in his heart he was a private enterprise. He used communism, the communist, like he used many other people—because he needed money—he needed support—he needed help etc., which he got from the communist party, but I don't think that he was ever really a socialist.[78]

One might add—nor perhaps ever really a multiculturalist. His actions suggest that at least after independence he adopted this position only when it was politically expedient. There was a profound estrangement between politicians who had fought for a common cause, evident in Kenyatta's manoeuvres to prevent Indians from joining high positions in government, and in Murumbi's frustration and resignation in December 1966.[79] This silencing of the multiplicity of voices affected Kenyan Indians and Kenyan Africans alike—though it was probably more noticeable for Kenyan Indians. Although Kenyatta had emphasized the connection and link between Africans and Indians at the time of the freedom movement and particularly, in the 1950s, once independence was achieved, such connections were disavowed.

Conclusion

This paper has illuminated the rise and fall of transimperial and transnational connections in the global south, and the vital role of the Indian diaspora in creating, nurturing, and maintaining these connections when they were at their peak in the 1940s, 1950s and 1960s. Five phases in these connections can be discerned: first, the pre-World War I era, in which tentative connections between India and East Africa were established, exemplified by the creation of the East African Indian National Congress modelled along the lines of the Indian National Congress. In a second, inter-war phase, transimperial connections between India and East Africa intensified, particularly in the exchange of ideas. The

78 Interview with Alberto, UK, May 2005.

79 Pinto's assassination and the increasing distance that Kenyatta created to Odinga have been interpreted as an oppression of left views. Good: Kenyatta and KANU, pp. 135–136; Rok Ajulu: The Road to Democracy. In: *Review of African Political Economy* 53 (1992), pp. 79–87; The Changing Face of Kenya Politics. In: *Transition* 25 (1966), pp. 44–50. Pinto's real murderer has never been convicted; it has been argued that the reason behind this is that his assassination was presumably initiated from high up in the government possibly with the involvement of the Directorate of Intelligence (Special Branch). See also Lefèbvre: The Lorenzo Natali Prize.

Second World War denoted a quieter period due to restricted communication and transport facilities during the war. The fourth phase, after World War II, saw a rapid take-off in transimperial and transnational connections between the newly independent India and the still imperially controlled East Africa. This period marks the most intensive links between India and East Africa, on an intellectual, social, and political level. After Kenyan independence, the political class in Kenya gave the highest priority to the process of building the nation-state. At the same time, the interest in transnational activities dwindled away: once *swaraj* and *uhuru* were achieved, the tendency to look inwards took hold in Kenya, denoting a growing distance from transnational politics towards strategies based on racialization and ethnic factionalism—in contrast to Tanzania and Ghana, where elements of transnationalism were kept up to a much larger degree under Nyerere and Nkrumah. In Kenya in the 1960s and the following decades, the narrative of a racial African nationalism came to replace a more secular, anti-imperial version of nationalism. 'Freedom fighters' who did not fit the picture were silenced or edited out, and not commemorated publicly.

This paper has illustrated that nationalism existed as an idea and as a practice. In the Kenyan context, transimperial and transnational relationships and connections, particularly through south-south exchanges, played crucial roles in the nationalist movement. In the practice of nationalism, or in the nationalist movement, in the pre-Independence period, the ideological and material input of South Asians was substantial. But in the process and practice of building a nation-state in post-independence Kenya, transnational connections were of reduced significance. The story of South Asian politicians and their involvement in Kenyan politics in the 1940s, 1950s, and 1960s, then, is one of a rise and fall of transnational politics and shows the ambiguities of these politics. Whereas the 1940s and 1950s saw a comparative openness to interaction and cooperation between African Kenyans and Indian Kenyans, the 1960s were characterized by a relative closure and increasingly exclusionary politics by the newly established Kenyan government.

In the 2000s, however, there was renewed interest in painting a more inclusive picture shaping national memory in a different way; hence the honours that have commemorated personalities such as Murumbi and Pinto in recent years. Why would this be the case? Perhaps a search in the past for the future is taking place: an attempt to return to the discourse of universal human rights in a transnational, or today, a global setting. This is clear from the text that accompanied the Milele Lifetime Achievement Award by the Kenya National Commission on Human Rights for Pinto in 2007, which refers to him as a Kenyan (not mentioning his Goan origins):

> Pio Gama Pinto was a Kenyan who fought for freedom against the colonialists, and later for justice against the Kenyan government when it failed to live up to its promises. In his short life, he contributed more to Kenya and civil rights than most of us could, given twice the lifetime. He believed in the equality of all Kenyans regardless of tribe or race.[80]

It also hints at a representation of the complex and intricate participation of a few, yet prominent, South Asians in the Kenyan nationalist movement. The process of analysing Kenya's past has undergone a transformation, amongst politicians as well as within the wider public. Attempts are being made in official politics to arrive at a more inclusive way of commemorating the so-called 'freedom fighters', reflecting a different way of thinking about the ambiguous relationship between African and Asian communities in Kenya and its representation in the public sphere. Moreover, these attempts reflect a greater distance of the public from the dominant hagiographic nationalist narratives. But perhaps more importantly, it reveals a greater preparedness to see nationalism in East Africa not so much in racialized terms as in its wider transnational context.

Acknowledgements

I am grateful to the University of Oxford Beit Fund, the University of Oxford History Faculty, OReNGA (Oxford Research Network on Government in Africa), the British Institute in Eastern Africa, the Alexander von Humboldt Foundation, the German Research Foundation and the Gerda Henkel Foundation for their financial support to the research on which this article is based. The Institute in Advanced Studies for the Humanities, University of Edinburgh, provided a congenial intellectual atmosphere in which to revise this article. Thanks also to Rana Mitter, Matthew Hilton and two anonymous reviewers

80 KNCHR: *4th Annual Human Rights and Democracy Awards Report*, p. 6.

for comments on an earlier version. Most importantly, though, I wish to thank all those who so generously shared their stories and experiences with me.

Bibliography

Ajulu, Rok: The Road to Democracy. In: *Review of African Political Economy* 53 (1992), pp. 79–87.

Allocation of Murumbi Grave a National Shame. In: *The Standard*, November 11, 2006.

Anderson, Benedict: *Imagined Communities: Reflections on the Origin and Spread of Nationalism*. London: Verso 2006.

Aminzade, Ronald: The Politics of Race and Nation: Citizenship and Africanization in Tanganyika. In: *Political Power and Social Theory* 14 (2000), pp. 53–90.

Anderson, David: *Histories of the Hanged. Britain's Dirty War in Kenya and the End of Empire*. London: Norton 2005.

Assmann, Aleida: *Erinnerungsräume: Formen und Wandlungen des kulturellen Gedächtnisses*. München: C. H. Beck 2010.

Berman, Bruce / John Lonsdale: *Unhappy Valley. Conflict in Kenya and Africa*, 2 vols. Athens / Ohio: Ohio University Press 1992.

Bharati, Agehananda: *Asians in East Africa: Jayhind and Uhuru*. Chicago: Burnham 1972.

Branch, Daniel: *Defeating Mau Mau, Creating Kenya: Counterinsurgency, Civil War, and Decolonization* (= African Studies, vol. 111). Cambridge / New York: Cambridge University Press 2009.

Branch, Daniel / Nicholas Cheeseman: The Politics of Controlin Kenya: Understanding the Bureaucratic-Executive State. In: *Review of African Political Economy* 33,107 (2006), pp. 11–31.

Brennan, James R.: South Asian Nationalism in an East African Context: The Case of Tanganyika, 1914–1956. In: *Comparative Studies of South Asia, Africa and the Middle East* 29,2 (1999), pp. 24–39.

The Changing Face of Kenya Politics. In: *Transition* 25 (1966), pp. 44–50.

Colonial Times, October 12, 1946. PRO, KV 2 / 1788.

Cooper, Frederick: Reconstructing Empire in British and French Africa. In: *Past & Present Supplement* 6 (2011), pp. 196-210.

Darwin, John: *The Empire Project. The Rise and Fall of the British World-System, 1830–1970*. Cambridge: Cambridge University Press 2009, pp. 514–565.

East African Standard, June 12, 1961.

Elkins, Caroline: *Britain's Gulag. The Brutal End of Empire in Kenya*. London: Pimlico 2005.

Gallery Befitting Murumbi. In: *The East African*, February 6, 2007.

Glassman, Jonathon: Sorting out the Tribes: The Creation of Racial Identities in Colonial Zanzibar's Newspaper Wars. In: *The Journal of African History* 41,3 (2000), pp. 395–428.

Good, Kenneth: Kenyatta and the Organization of KANU. In: *Canadian Journal of African Studies* 2,2 (1968), pp. 115–136.

Gregory, Robert G.: *India and East Africa: A History of Race Relations within the British Empire, 1890–1939*. Oxford: Oxford University Press 1971.

—: *South Asians in East Africa: An Economic and Social History 1890–1980*. Boulder: Westview Press 1993.

—: *Quest for Equality. Asian Politics in East Africa 1900–1967*. Hyderabad: Orient Blackswan 1993.

Gupta, Akhil: The Song of the Non-aligned World: Transnational Identities and the Reinscription of Space in Late Capitalism. In: *Cultural Anthropology* 7,1 (1997), pp. 63–79.

Herzig, Pascale: *South Asians in Kenya: Gender, Generation and Changing Identities in Diaspora*. Münster: Lit 2007.

Himbara, David: Failed Africanization of Commerce and Industry. In: *World Development* 22,3 (1994), pp. 469–482.

Hornsby, Charles: Social Structure of the National Assembly in Kenya, 1963–1983. In: *The Journal of Modern African Studies* 27,2 (1989), pp. 275–296.

Kago, Tony: Alarm over Planto Grab the Grave of Former VP. In: *The Nation*, January 18, 2007.

Kenya National Commission on Human Rights (KNCHR): *4th Annual Human Rights and Democracy Awards Report* (March 26, 2007).

Kyle, Keith: Gandhi, Harry Thuku, and Early Kenya Nationalism. In: *Transition* 4,27 (1966), pp. 16–22.

Lefèbvre, François: The Lorenzo Natali Prize: Human Rights and Development, the Real Winners. In: *The Courier ACP-EU* 190 (January–February 2002), p. 30.

Lonsdale, John: KAU's Cultures: Imaginations of Community and Constructions of Leadership in Kenya after the Second World War. In: *Journal of African Cultural Studies* 13,1 (2000), pp. 107–124.

Maciel, Mervyn: *Bwana Karani*. London: Merlin 1985.

Madhvani, Manubhai: *Tide of Fortune: A Family Tale*. Noida: Random House 2009.

The Making of a Nation. A Political History of Kenya (Kenya, 2008 D: Hilary Ngweno).

Mamdani, Mahmood: Beyond Settler and Native as Political Identities: Overcoming the Political Legacy of Colonialism. In: *Comparative Studies in Society and History* 43,3 (2001), pp. 651–664.

Manela, Erez: *The Wilsonian Moment. Self-Determination and the International Origins of Anticolonial Nationalism*. New York: Oxford University Press 2007.

Mangat, Jagjit S.: *A History of the Asians in East Africa c.1886 to 1945*. Oxford: Oxford University Press, 1969.

Mathangi, Patrick: Murumbi Burial Site Grabbed, Claims Trust. In: *The Standard*, November 9, 2006.

Mkenda, Festo: *Building National Unity in Sub-Sahara Africa: The Impact of State Policies on the Chagga Community in Northern Tanzania*. University of Oxford DPhil thesis. Oxford 2009.

Morris, Harold S.: *The Indians in Uganda: Caste and Sect in a Plural Society*. London: Weidenfeld & Nicolson 1968.

Murumbi's Grave Now Protected. In: *The Nation*, January 20, 2007.

Mutiga, Muriithi: Nationalist Murumbi's Grave Vandalised. In: *Sunday Standard*, May 1, 2005.

Nagar, Richa: The Politics of Gendered Boundaries: South Asian Communities in Tanzania. In: *Ghadar* 1,1 (1997), pp. 6–7.

—: Saboteurs? Or Saviours? The Position of Tanzanian Asians. In: *Samar* 13 (2001), pp. 14–19.

Ngotho, Kamau, Kenya: How Pinto Murder Was Plotted … And Kisilu Framed. In: *The Daily Nation*, June 19, 2000.

—: Kenya: The Pinto Murder Mystery Resolved. In: *The Daily Nation*, June 18, 2000.

Nora, Pierre: *Les lieux de mémoire*. Paris: Gallimard 1984.

Nowrojee, Pheroze: *Pio Gama Pinto. Patriot for Social Justice*. Nairobi: Sasa Sema Publications 2007.

Oonk, Gijsbert: *Asians in East Africa: Images, Histories & Portraits*. Arkel: SCA Producties 2004.

Patel, Ambu: *Struggle for Release Jomo & His Colleagues*. Nairobi: New Kenya Publishers 1963.

Patel, Zarina: *Challenge to Colonialism. The Struggle of Alibhai Mulla Jeevanjee for Equal Rights in Kenya*. Nairobi: Publishers Distribution Services 1997.

—: *Unquiet. The Life and Times of Makhan Singh*. Nairobi: Zand Graphics 2006.

Odinga, Oginga: *Not Yet Uhuru. The Autobiography of Oginga Odinga*. London: Heinemann 1967.

Odinga, Oginga et al.: *Independent Kenya's First Martyr. Socialist and Freedom Fighter – Pio Gama Pinto*. Nairobi: Pan Africa Press 1966.

Salvadori, Cynthia: *Through Open Doors: A View of Asian Cultures in Kenya*. Nairobi: East African Educational Publishers 1989.

Schubert, Frank: Die kolonialen Wurzeln nachkolonialer Gewalt: Die Entstehung von Martial Tribes und politischer Ethnizität in Uganda. In: *Archiv für Sozialgeschichte* 48 (2008), pp. 275–294.

Seidenberg, Dana April: *Uhuru and the Kenya Asians: The Role of a Minority Community in Kenya Politics, 1939–1969*. New Delhi: Vikas 1983.

Singh, Makhan: *History of Kenya's Trade Union Movement to 1952*. Nairobi: East African Pub. House 1969.

—: *Kenya's Trade Unions, 1952–1956*. Nairobi: Uzima Press 1980.

Subrahmanyam, Sanjay: Connected Histories: Notes towards a Reconfiguration of Early Modern Eurasia. In: *Modern Asian Studies* 31,3 (1997), pp. 735–762.

Thatiah, Peter: Collector par Excellence. In: *The Standard*, November 18, 2006.

Thurston, Anne: Emergence of a Nationalist: An Interview with Joseph Murumbi, part II. In: *Kenya Past and Present* 11 (1979), pp. 23–28.

—: Formation of a Character: An Interview with Joseph Murumbi, part I. In: *Kenya Past and Present* 11 (1979), pp. 16–21.

Warah, Rasna: *Tripple Heritage: A Journey to Self-Discovery*. Nairobi: Warah 1998.

II.
Narrating the Ocean

Africa's Asian Options
Indian Ocean Imaginaries in East African Literature

Frank Schulze-Engler

At least since the beginning of the new millennium, African-Asian relations have not only become a major hotspot in global mediascapes, but have also increasingly caught the attention of African and Asian studies. If during the 1980s and '90s a lot of scholarly effort went into reconstructing the long history of Asian-African contact across the Indian Ocean, current interest seems to focus on the significance of African-Asian interactions for the global dynamics of the twenty-first century present and on the potential of these interactions for the future of economic, political, social and cultural relations worldwide. Together with other transcontinental interactions within what used to be called the "Global South" (such as those between Africa and Latin America), African-Asian relations are often perceived as signalling a transition in global affairs from a "Eurocentric"—or more precisely "Western-dominated"—world order to a "multipolar" or "post-American"[1] world, in which older power constellations are still present, but have lost their capacity to define the whole system and are increasingly supplemented by a plurality of new alignments.
It is in this context that "Africa's Asian Options" have received renewed attention in a number of academic disciplines in recent years.[2]

1 See Isabel Hofmeyr: Universalizing the Indian Ocean. In: *Publications of the Modern Language Association of America* 125,3 (2010), pp. 721–729, here p. 721: "As an arena in which these developments intersect, the Indian Ocean offers a privileged vantage point from which to track a changing world order. The view from the Indian Ocean permits us to look back to the lingering effects of the cold war and forward to what some are calling a 'post-American' world."

2 This paper has emerged from research conducted in the framework of "Africa's

Given the sheer volume and visibility of China's investments in Africa, much of this attention has focused on China's role in Africa and on the significance of economic relations in African-Asian interactions. Yet these interactions involve numerous other Asian actors (such as India, Japan, Korea, Malaysia, the Philippines, Taiwan or Vietnam) and they also extend to other fields such as politics, civil society, migration and culture. Against the background of these larger dynamics, cultural and literary studies are thus faced with the intriguing question to what extent new social imaginaries are forming alongside and as part of Africa's Asian options: if current African-Asian interactions constitute more than an ephemeral episode, they are likely to leave their traces in (and in turn to be shaped by) a cultural undertow involving conflicts, negotiations and realignments.

The present essay presents selected examples of such conflicts, negotiations and realignments in Anglophone East African writing. It situates itself within Indian Ocean studies, which have gained new prominence in cultural and literary studies in recent years, and thus begins with a brief overview of the insights and challenges generated by Indian Ocean studies. Taking its cue from recent critical self-reflexions in the field that have highlighted an "Indocentrism" in Indian Ocean research on the one hand and a tendency towards romanticising Indian-African interactions on the other, the main part of this essay presents examples from East African literature relating to Indian Ocean imaginaries beyond India and to an unflinching scrutiny of African-Indian political, social and cultural relationships within a field of tension constituted by the polarities of diaspora and citizenship.

1. Indian Ocean Studies: Insights and Challenges

One of the key insights of contemporary Indian Ocean studies lies in the acknowledgement of the seminal role of the larger Indian Ocean region in twenty-first century globalisation processes:

> It is my contention that the Greater Indian Ocean, stretching eastward from the Horn of Africa past the Arabian peninsula, the Iranian plateau, and the Indian

Asian Options: Frankfurt Inter-Centre-Programme on new African-Asian Interactions" (AFRASO), a major new interdisciplinary collaborative project at Goethe University Frankfurt funded by the German Federal Ministry of Education and Research (BMBF) that brings together some 40 researchers from 6 faculties. For further information see http://www.afraso.org (accessed February 27, 2014).

> Subcontinent, all the way to the Indonesian archipelago and beyond, may comprise a map as iconic to the new century as Europe was to the last one. Hopefully, the twenty-first century will not be as violent as the twentieth, but, to a similar degree, it could have a recognizable geography. In this rimland of Eurasia – the maritime oikoumene of the medieval Muslim world that was never far from China's gaze – we can locate the tense dialogue between Western and Islamic civilizations, the ganglia of global energy routes, and the quiet, seemingly inexorable rise of India and China over land and sea.[3]

"The future of American power" and the need to redefine American strategic interests with regard to political alliances and energy supplies are not the only concerns, however, that have emerged from the study of the 'Greater Indian Ocean' region. Indian Ocean studies have also raised important issues relating to a fundamental reassessment of economic, political and social relations on a global scale:

> But how European is the global system today – or, for that matter, in earlier ages when the Indian Ocean world did rather well for itself without European dominance? In this volume we turn to the Indian Ocean world as an intriguing sphere of mutual exchange on economic and cultural levels that continues to run parallel to the current dominance of the West: in a word, we are looking at alternate universalisms, analogous to the discourse communities that hold together Latin American cultures, Eastern European, Western Asian, Southeast Asian histories – "global" histories defined by preoccupations other than those of the West.[4]

This reassessment is particularly important in the case of Africa, since both globalisation studies and African studies have tended to regard Africa as a backwater of globalisation:

> My second aim […] is to challenge the dominant reading of Africa as having somehow escaped or bypassed the processes of globalization. This reading echoes the empty historical slot assigned to Africa most famously by Hegel and disserves the complex negotiations of commerce, identity, and the advent of modernity performed by a multitude of actors on the African stage. Indeed, along with an increasing number of my Africanist colleagues, I argue that African globalization must be thought of in the context of a long history. The role of Africa in the Indian Ocean trade system is crucial to interrogating established readings of Africa.[5]

3 Robert D. Kaplan: *Monsoon: The Indian Ocean and the Future of American Power.* New York: Random House 2010, p. xi.

4 John C. Hawley: Introduction: Unrecorded Lives. In: John C. Hawley (ed.): *India in Africa, Africa in India: Indian Ocean Cosmopolitanisms.* Bloomington / Indianapolis: Indiana University Press 2008, pp. 1–15, here p. 10.

5 Gaurav Desai: Oceans Connect: The Indian Ocean and African Identities. In: *Publications of the Modern Language Association of America* 125,3 (2010), pp. 713–720, here p. 715.

This emphasis on countering the invisibility of Africa in current accounts of globalisation has also surfaced in cultural and literary studies, where critics have perceived Indian Ocean studies as a chance to revise long-established narratives of Africa's relations to the wider world based on an 'Atlanticist' perspective on slavery and on the legacies of the 'Middle Passage', which need to be supplemented by an understanding of Africa's 'Eastern transnationalism' based on a very different history of slavery on the one hand and transoceanic trade relations on the other:

> As transnationalism and associated models like oceanic studies gain momentum in the academy, the Indian Ocean obliges us to extend our axes of investigation. It requires us to relativize the Atlantic, which has become normative, especially in slavery and African diaspora studies. Analyses of domination and resistance, colonizer and colonized arise from postindependence revisions of colonial history. These in turn draw on Atlantic readings of slavery where the boundary between enslaved and free is unequivocally demarcated and racialized. In the Indian Ocean this boundary blurred, and race and slavery were not associated in any marked form [...]. At every turn the Indian Ocean complicates binaries, moving us away from the simplicities of the resistant local and the dominating global and towards a historically deep archive of competing universalisms.[6]

Indian Ocean studies have thus emerged as a complex, multidisciplinary field with repercussions in a large number of disciplines:

> The Indian Ocean is a theoretical terrain, a geographical space and a historical network of human connectivities. It is by its nature (and culture) a fluid topic, and one of ongoing interdisciplinary interest – it has been a focus of study for historians, anthropologists, literary scholars and cultural studies critics alike. Scholars have neither exhausted the potential of looking at comparative models such as the Black Atlantic and the Mediterranean for possible insights into the specificities of the Indian Ocean nor stopped probing the surfaces, depths and rims of the Indian Ocean waters to unearth previously unwritten histories of transnationalism, migration and diasporic imaginings.[7]

Although Indian Ocean studies in this wider, interdisciplinary sense are still a relatively new field of enquiry, they have already begun to

6 Isabel Hofmeyr: Universalizing the Indian Ocean. In: *Publications of the Modern Language Association of America* 125,3 (2010), pp. 721–729, here pp. 721–722. For a critique of "Middle Passage epistemology" see also Michelle M. Wright: Middle Passage Blackness and its Diasporic Discontents: The Case for a Post-War Epistemology. In: Eve Rosenhaft / Robbie Aitken (eds): *Africa in Europe: Studies in Transnational Practice in the Long Twentieth Century*. Liverpool: Liverpool University Press, 2013, pp. 217–233.

7 Pamila Gupta: Introduction. In: Ead. / Isabel Hofmeyr / Michael Pearson (eds): *Eyes Across the Water: Navigating the Indian Ocean*. Pretoria: Unisa 2010, pp. 3–4, here p. 4.

self-reflexively scrutinise their theoretical presuppositions and methodological assumptions. In a recent essay entitled "Africa as a Fault Line in the Indian Ocean", Isabel Hofmeyr offers a critical overview on current trends in Indian Ocean studies and argues that one major fault-line in contemporary Indian Ocean studies is constituted by "the prominence of India as opposed to the invisibility of Africa".[8] Her argument in this respect is based on the historiography of Indian diasporic populations in South Africa and in East Africa, and on the fact that this historiography has made up a great proportion of what has been considered "Indian Ocean studies" so far. A second critique focuses on the theoretical premises on which much of Indian Ocean studies have relied. According to Hofmeyr, there is an "implicit Third Worldism" at work in Indian Ocean studies that has largely remained un-reflected and thus constitutes an obstacle on the way towards a more nuanced understanding of contemporary Indian Ocean realities:

> [The Indian Ocean] will assume an added significance in shaping the new post-Cold War world order. Yet, within the academy we are currently poorly prepared to explain or narrate this emerging moment. The existing models that we have (anti-colonialism, post-colonial studies, non-alignment, south-south, area studies) are ill-equipped to help us understand the new politics and rivalries of the Indian Ocean. One might of course expect that Indian Ocean Studies would come to the rescue, but this field of endeavor has equally been inflected by existing models of anti-colonialism and south-south solidarity. […] [O]ne strand in Indian Ocean Studies arises from erstwhile area studies scholars who follow their subjects to sea, taking the implicit Third Worldism of area studies with them. […]
> In order to think about how Indian Ocean Studies might be revised in order to address the complexities of the new era, the implicit Third Worldism needs to be surfaced and made visible, not in order to abandon it but rather to make it more complicated and hence able to speak to the complexities of the emerging moment.[9]

The crypto-normative 'Third Worldist' values underlying academic practice, Hofmeyr argues, have also influenced the manner in which Indian Ocean studies have addressed Indian-African relations in the colonial and post-independence era. On the one hand, implicit ideals of 'South-South' solidarity may have caused academics to bypass the

8 Isabel Hofmeyr: Africa as a Fault Line in the Indian Ocean. In: Ead. / Pamila Gupta / Michael Pearson (eds): *Eyes Across the Water: Navigating the Indian Ocean*. Pretoria: Unisa 2010, pp. 99–108, here p. 101.
9 Ibid., pp. 100–101.

tensions between Indians and Africans and to focus research on clearly demarcated ethnic communities:

> These uneasy histories of African-Indian friction remain largely unexplored and are seldom discussed in public. Segregationist and apartheid structures in South (and East) Africa exacerbated these silences while academic research reproduced ethnic boundaries tending to focus on either 'Indian' or 'African' communities.[10]

On the other hand, accounts of African-Asian interaction (for example in the context of anticolonial movements in East and Southern Africa) have often highlighted political solidarity, but have seldom connected these accounts to the contested terrain of everyday interactions and identifications:

> [W]hile such studies of African-Indian solidarity certainly exist, they are seldom offset against histories of African-Indian conflict. This division produces a bipolar historiography where there is either too much sentimental shoulder-to-shoulder solidarity or too little. In short, the historiography veers between romantic stereotype or silence.[11]

Both strands of Hofmeyr's critique (the preponderance of India and the soft-pedalling of African-Asian interactions) undoubtedly highlight important challenges that Indian Ocean studies need to come to terms with in order to grasp the full complexity of their subject. As this essay hopes to show, East African literature has already begun to face up to these challenges, both by creating Indian Ocean imaginaries beyond the Africa-India nexus and by scrutinising the emergence of contested Indian-African identities posited in a field of tension constituted by diasporic allegiances on the one hand and Indian struggles for African citizenship on the other.

2. Indian Ocean Imaginaries: Beyond India

An intriguing example of a perception of East Africa in terms of a larger Indian Ocean region can be found in an essay on the Tamarind Market in Mogadiscio by Nuruddin Farah. For Farah, the expulsion and flight of many of Mogadiscio's citizens and the (partial) resettlement of the city by non-urban settlers from other parts of Somalia that changed the face of the city in the wake of the collapse of the

10 Hofmeyr: Africa as a Fault Line in the Indian Ocean, p. 101.

11 Ibid., p. 102.

Somalian nation state signals the victory of a 'landlocked' notion of Africa and the destruction of a cosmopolitan spirit embodied in the Tamarind Market, an economic and social hub that integrated Mogadiscio into a wider Indian Ocean world:

> The history of Mogadiscio, how it came into being and what became of it after it went up in flames following the civil war, are to my mind all tied up with the history and destiny of the small cosmopolitan community who ran the Tamarind Market. The presence of this small community dates back to the tenth century, at which time Mogadiscio existed as a city state and boasted a negligible level of administration run for the benefit of the bourgeois elite, many of whom came from elsewhere: Iran, India and Arabia. As more and more foreigners migrated to it from other countries over the years, the city assumed an unmistakably cosmopolitan orientation. It was an open city with no walls, to which anyone could come, provided he or she lived in harmony and at peace with those already there. It was as small as many other cities in other parts of the world then, probably no bigger than four square kilometres. And it was prosperous, thanks to its residents, many of them artisans hailing originally from the Middle East or the Indian subcontinent. [...]
> Perhaps what the warlords and their irregular armies managed to destroy in Somalia was not the infrastructure of the city, of which there was very little, or the foundation of the state, of which there was hardly any worth saving. Rather, they destroyed the spirit of a place like Tamarind Market, murdering the people who ran it, chasing out those who frequented it, in short demolishing the idea of cosmopolitanism.[12]

While Farah's juxtaposition of a 'terrestrial' and a 'transoceanic' vision of Africa takes its cue from recent developments in war-torn Somalia, Abdulrazak Gurnah in his novel *By the Sea* returns to the era of British colonialism, when the Indian Ocean was infamously regarded as a 'British Lake', to develop an Indian Ocean imaginary encompassing not only the riparian nations around the Arabian Sea, but embedding East Africa in a wider African-Asian trans-region stretching right up to China. One of the protagonists of the novel recalls a school lesson in colonial Zanzibar that centrally featured a map drawn by the (British) teacher to locate Zanzibar in the wider world:

> As his story developed, he began to draw a map on the blackboard with a piece of white chalk: the coast of north-west Europe, the Iberian peninsula, southern Europe, the land of Shams, Syria and Palestine, the coast of North Africa which then bulged out and tucked in and then slid down to the Cape of Good Hope.

12 Nuruddin Farah: Of Tamarind & Cosmopolitanism. In: Ntone Edjabe / Edgar Pieterse (eds): *African Cities Reader*. Johannesburg: African Centre for Cities / Chimurenga Magazine 2010, pp. 9–12, here pp. 10–12.

> As he drew, he spoke, naming places, sometimes in full sometimes in passing. Sinuously north to the jut of the Ruvuma delta, the cusp of our stretch of the coast, the Horn of Africa, then the Red Sea coast to Suez, the Arabian peninsula, the Persian Gulf, India, the Malay Peninsula and then all the way to China. He stopped there and smiled, having drawn half the known world in one continuous line with his piece of chalk. He put a dot halfway down the coast of East Africa and said, 'This is where we are, a long way from China.'[13]

The stunning last sentence in this passage can arguably be read in at least two different ways. The context for the teacher's memorable performance is the story of Christopher Columbus's 'discovery' of the new World, and the amazing map the teacher sketches onto the blackboard is thus ultimately designed to make the pupils in Zanzibar understand the global dimension of Columbus's westbound quest for the sea-route to China. "This is where we are, a long way from China" might thus be read as part of an imperial mapping project designed to overawe East African 'locals' by the global scope of Europe's territorial ambitions, with Zanzibar reduced to an insignificant spot on the world map. There are good reasons to refute such a reading, however, and to see the last sentence of the above passage as pointing into a quite different direction.

Gurnah's novel is full of accounts of journeys across the Indian Ocean and, in fact, presents the history of East African coastal cities in general and of Zanzibar in particular as embedded in long-distance economic, social and cultural networks linking East Africa to Arabia as well as South Asia. The teacher's fortuitous offering of a sense of place relating the colonial East African pupil to Asia ("This is where we are, a long way from China" rather than "This is where we are, a long way from Britain") can thus be read in conjunction with the various accounts of trans-Indian Ocean relationships in Gurnah's novel as an epiphanic glimpse of a wider trans-regional vision extending beyond the long-standing aquatic links between East Africa and South Asia generally considered to be at the heart of "Indian Ocean Imaginaries".

Yet the Indian Ocean imaginaries alluded to in *By the Sea* are diverse and cannot simply be associated with the image of a peaceful world of trade, contact and multiplicity that has often been generated in Indian Ocean studies, particularly with regard to the precolonial era. East

13 Abdulrazak Gurnah: *By the Sea*. London: Bloomsbury 2002 [2001], p. 37.

African contact with the Arab world also generated a wide array of (often antagonistic and violent) relations based on hegemony and discrimination that left powerful legacies reaching into and even beyond the colonial era:

> For centuries, intrepid traders and sailors, most of them barbarous and poor no doubt, made the annual journey to that stretch of coast on the eastern side of the continent, which had cusped so long ago to receive the musim winds. They brought with them their goods and their God and their way of looking at the world, their stories and their songs and prayers, and just a glimpse of the learning which was the jewel of their endeavours. And they brought their hungers and greeds, their fantasies and lies and hatreds, leaving some among their numbers behind for whole life-times and taking what they could buy, trade or snatch away with them, including people they bought or kidnapped and sold into labour and degradation in their own lands. After all that time, the people who lived on that coast hardly knew who they were, but knew enough to cling to what made them different from those they despised, among themselves as well as among the outlying progeny of the human race in the interior of the continent. […]
> New maps were made, complete maps, so that every inch was accounted for, and everyone now knew who they were, or at least who they belonged to. Those maps, how they transformed everything. And so it came to pass that in time those scattered little towns by the sea along the African coast found themselves part of huge territories stretching for hundreds of miles into the interior, teeming with people they had thought beneath them, and who when the time came promptly returned the favour.[14]

The third example comes from a collection of oral stories entitled *Eating Chiefs: Lwo Culture from Lolwe to Malkal Selected, Interpreted and Transmuted by Taban lo Liyong*, published in 1970. One of the stories that Taban lo Liyong, one of the stalwarts of East African Literature in English, presents in this collection dedicated to exploring intriguing interfaces of oral and literary culture, is a 'classical' genealogical tale that he retitled "The Spear, Bead and Bean Story". The story that is meant to explain the division of the Luo into different groupings revolves around two brothers who used to live together with their families, but later separated because of a feud between the brothers. The feud begins when one of the brothers borrows a spear from the other, but loses this spear in an elephant hunt because the wounded elephant runs off with it. The other brother demands his spear back and forces the hapless hunter to recover the spear, which he does, risking his life and returning badly mauled. A short time later, the daughter of the

14 Ibid., pp. 15–16.

other brother plays with beads borrowed from the erstwhile hunter and swallows one of these beads. The other brother now insists on his property rights and demands his bead back; the child's belly is cut open, the bead is retrieved and the two brothers and their clans from then on go their separate ways.

In his rendering of this story, Taban lo Liyong turns the tragic death of the child into an act of creation: the opening of the child's belly and the recovery of the bead signal the birth of a syncretism that according to lo Liyong is an inevitable feature of allegedly 'traditional' African cultures. The bead, a central symbol of Luo royal power, no longer stands for the internal division of an allegedly 'pure' culture, but for the links that bind that culture to other cultures and locations—locations as far away as China:

> Royalty wears strange articles, especially Lwo royalties who have accumulated many strange subcultures in their wanderings. One of these is a string of beads, blue beads. It strikes me as odd that a slit-eyed Chinese, labouring away under a stern mandarin, makes some blue beads for Chinese populace or peasants, and that these beads wend their way through all Asia, through north-east Africa, and finally this product becomes adopted by Lwo royalty for kings. [...] Also that the blue bead should become a culture symbol, rivalling the spear – the spear, being itself an import, a foreign substance originating in far-away Merowe, or wherever it was, makes me say that when we speak of a 'pure' culture, we mean foreign amalgams nationalized. [...]
> If the spear symbolizes brute force such as was Achille's, the adoption of the far-travelled foreign blue bead, the beautifying article of culture-absorption, symbolizes the intellectual power for openness to the non-material, the progressive, and the new and foreign, among the Lwo.[15]

While lo Liyong's imaginary here is decidedly non-aquatic (his account of the bead's journey from China to Africa in fact suggests the land route via Central and Western Asia and Arabia rather than the classic Indian Ocean crossing), it can arguably be seen as an example of an extended Indian Ocean Imaginary that links up (in this case, oral and allegedly 'traditional') African culture with Asia.

15 Taban lo Liyong: The Spear, Bead and Bean Story. In: Id. (ed.): *Eating Chiefs: Lwo Culture from Lolwe to Malkal Selected, Interpreted and Transmuted by Taban lo Liyong*. London: Heinemann 1970, pp. 3–6, here pp. 4–5.

3. Indian-African Writing: Between Diaspora and Citizenship

In an extensive study of Indian-South African Writing, Pallavi Rastogi has argued that in the field of tension created by Indian diasporic identity on the one hand and the desire to attain full citizenship in a new, post-Apartheid South Africa on the other, Indian South African writers have overwhelmingly opted for the assertion of citizenship. Indian-African writing, Rastogi suggests, thus does not belong to the field of 'diasporic' literature, but should be regarded as 'Afrindian' literature instead:

> Indians desire South African citizenship in the fullest sense of the word, a need for national anchorage that is a consequence of their erasure in both the apartheid and postapartheid consciousness. This longing for belonging is asserted through an "Afrindian" identity. The term suggests both an Africanization of Indian selfhood and an Indianization of South Africa. The former is achieved by an affiliation with the indigenous population and an attachment to the African land, while the latter is demonstrated through tracing the changes wrought in South Africa by the Indian presence. Changing oneself as well as being an agent of change in order to claim a South African national identity is the central dialectic underpinning Indian fiction in South Africa.[16]

Because negotiating one's allegiance to the new South African polity is so paramount in 'Afrindian' writing, Rastogi holds, diasporic concerns linking writers of Indian origin to their—imaginary—'mother country' have dwindled away, and 'India' has become an 'empty signifier' in contemporary writing:

> In the case of South African fiction, even though the "Mother Country" is an icon charged with mythic resonances, there is rarely a desire to return even in a "spiritual" sense. India, then, exists as an empty symbol. The proclamation of a South African national identity, as well as the articulation of a permanent bond with an everyday South Africa rather than with an imagined India, dominates the thematic concerns of South African Indian fiction. The critical genealogy of Indian writing traced in this book asserts, above all, the Indian allegiance to South Africa.[17]

Given the central hypothesis of her study, it is hardly surprising that Rastogi strictly delimits her analysis and presents her category of 'Afrindian' writing as coterminous with South African national space. Yet the central insights provided by her model of a field of tension

16 Pallavi Rastogi: *Afrindian Fictions: Diaspora, Race, and National Desire in South Africa.* Columbus: Ohio State University 2008, pp. 1–2.
17 Ibid., p. 10.

constituted by diasporic modes of consciousness linking Indo-African populations and cultures to an—often imaginary—Indian 'Homeland' on the one hand and the desire to safeguard 'Afrindian' citizenship rights and full civil participation in independent African nations on the other are clearly relevant beyond the South African national context and can arguably also be applied to East Africa. Instead of a strict focus on citizenship alone (which effectively dismisses any form of diasporic consciousness as a reactionary relic of the past), a more flexible approach seems to be called for in the East African context, however: the expulsion of some 75,000 Indians from Uganda by military dictator Idi Amin in 1972 as well as lower profile tensions between Africans and South Asians in other East African countries such as Kenya or Tanzania have resulted in a sociopolitical reality where the struggle for citizenship rights in African nation states remains on the agenda, but where transnational networks and diasporic connections (not only 'back' to India but also 'forward' to other diasporic locations such as North America or Europe) play a major role in political, social and cultural self-identifications—which in turn strongly affect Indian-East African literature.

While even in the South African context it seems implausible to delimit contemporary writing to a strictly national perspective at a time when African literary studies have long since begun to explore the transnational, transcultural, diasporic or 'Afropolitan'[18] dimensions of African literature, a 'methodological nationalism' restricting literary studies to an exclusively nation-bound perspective is bound to miss major dimensions of what constitutes 'Indian-African' (and other) writing in East Africa today. Rather than exclusively focusing on the 'deterritorialisation' of Indo-African writing (and thus confining it to a perennial state of diasporic consciousness down the generations) or the 're-territorialisation' of Afrindian literature in the epistemological space of Africa's post-independence nation states (and thus conceptually segregating it from the transnational or transcultural dynamics that characterises much of contemporary African literature), a trans-regional Indian Ocean perspective can arguably help to come to terms

18 See, for example, Jennifer Wawrzinek / J. K. S. Makokha (eds): *Negotiating Afropolitanism: Essays on Borders and Spaces in Contemporary African Literature and Folklore.* Amsterdam / New York: Rodopi 2011.

with the complex articulations and tensions that characterise literary representations of African-Asian interactions in modern African literature. 'Indian Ocean Imaginaries' may thus be at work even beyond the 'aquatic' dimensions of transoceanic travel, trade and exchange; political Indian Ocean imaginaries are arguably also constituted along an axis constituted between diasporic consciousness on the one hand and the struggle for citizenship on the other.

Peter Nazareth's *In a Brown Mantle* provides a fascinating example of the contradictory impulses and identity negotiations generated by diasporic consciousness and the struggle for citizenship in East African literature. Set in the transition years between late colonial rule and early independence in Damibia (a fictionalised East African nation bearing close resemblance to Uganda), the novel follows the transformation of Deogratius D'Souza, an East African Asian of Goan ancestry, from indifferent bystander to anticolonial militant in the nationalist movement against British rule. After independence D'Souza becomes a senior government official, but eventually loses out in the internal power-struggles that pave the way for the future military dictator of Damibia, and finally has to leave East Africa to go into exile in Britain.

The immersion of Asian Africans in the anticolonial movements in the mid-twentieth century and the Asian struggle for citizenship in the new African nation states created after independence thus figure prominently in Nazareth's novel. Yet the text does not present a one-dimensional image of Asian-African solidarity: many of the Goans (particularly the economically more successful and politically influential community leaders) are shown to have sided with British colonialism during the turbulent transition towards independence, and Deogratius D'Souza, the intra-diegetic narrator from whose perspective the story is told, is by no means an immaculate freedom fighter, but in many respects an unreliable narrator who at first keeps out of politics and is only later, somewhat reluctantly, drawn into the anticolonial movement headed by the charismatic Kyeyune, later to become the first president of independent Damibia.

Early on in the novel, the Goan community in Damibia is presented as a social group intent on keeping its social and cultural traditions intact. Its policy of maintaining its 'cultural heritage' in East Africa is supported by the colonial authorities who promote Asian ethnic identities in a bid to maintain ethnic differences and to forestall cross-ethnic political mobilisation among their subjects:

> The Abala Goan Institute was the oldest of the Institutes, having been built at the turn of the century. In those days, there were clear racial boundaries under colonial rule and each race had to turn inwards, so to speak, to make its own form of entertainment. The first group of Goans in Abala apparently felt the need for an institution to cater for their social, moral, intellectual, and educational needs. They obtained land from the Government to put up a building, permission no doubt being granted because it was in the interests of the Government to keep its faithful and efficient civil servants contented.[19]

This ethnically demarcated identity politics that goes hand in hand with a political conservatism with regard to the British colonial regime in East Africa is challenged, however, when the first stirrings of an increasingly radical anticolonial movement make their presence felt at the Abala Goan Institute. In a passage characteristically narrated by D'Souza from a bystander's perspective, Pius Cota, an anticolonial political activist with Goan roots from a neighbouring country (clearly recognisable as Kenya in the fictive political geography of Nazareth's text) disrupts the acquiescent consensus among the Goan social elite and confronts the community stakeholders with the reality of their position as colonial subjects within the colonial hierarchy:

> As I got to the fringe of the group, I heard him talking about "non-cooperation with the Government," "overthrowing the Government," "Independence," and "Africans ruling themselves." This was a strange fellow, I thought. He talked about the horrors committed by the colonial Government. He said that the black people in his country were treated like dogs, but they were beginning to resist, and a new freedom movement had started. The conclusion he drew was that it was necessary for the Goans of Damibia to throw in their lot with the cause for African Independence before there was guerilla fighting.
> "Nonsense," said Bernie Rodrigues, one of our senior members. [...]
> "Nonsense. Exploitation, indeed! Anybody can see the good, the grand and glorious progress brought about by the British. Before the British touched this country with their golden glove, Damibia was a mere swamp, a jungle, with a few mud huts, malaria, blackwater fever, and sleeping sickness. And now – now it is the veritable Jewel of Africa, the Apple of the Lord's eye. Schools, houses, offices, hospitals, shops, yea–"

19 Peter Nazareth: *In a Brown Mantle*. Nairobi: East African Literature Bureau 1972, p. 8.

> "Bullshit!" said Pius. "Who benefits? How many Africans can go to the hospitals? In my country, Blacks and Browns cannot go to any hotels. You there" – turning to one of the by-standers – "how many hotels can you go to?"[20]

While part of the Goan community accepts its status as 'Asian' colonial subjects in a tiered system where they are relatively privileged compared of the African population, others join the anticolonial liberation struggle and begin to develop a new, non-racial or non-ethnic notion of African citizenship. Yet even Deogratius D'Souza, who eventually becomes one of the organising brains behind the anticolonial movement in Damibia and decisively breaks with an 'ethnic' consciousness that would consider Asians as superior and Africans as inferior, at times succumbs to a possibly racist, but certainly highly exoticist perception of 'Africa'. In a striking passage employing free indirect speech, the narrative follows D'Souza into a nightclub which turns into a veritable hothouse of gendered stereotypes all too familiar from the multiple European discourses of 'colonial desire':[21]

> If you were from outside Eastern Africa, you would think of the New Jazz Nightclub as just another nightclub, though rather exotic. To a Goan, it was an *African* nightclub carrying with it overtones of the unusual, the underworld, possibly even a tinge of the depraved. [...]
> The hall is dark and full of music. There is movement everywhere, people dancing or drinking. They are dressed as they like and do what they like. The atmosphere is charged, as though the uninhabited human being is inescapably erotic. Or is it that one is attracted by seeing so many dark, unaccompanied women? [...]
> You shake your hands and your legs, flinging everything all over the place. Your African partner is already in a world of her own, her face bearing an expression of not-knowingness. And the vibrations radiate from your stomach until your legs, your genitals, your chest, your brain – all are one mass of electronic rhythm. And the music swells and throbs. Time ceases to matter. All becomes a seething mass, like sperm in a release of semen.
> [...]
> You are already part of the whole pulsating universe. The whole electric universe. Until the partner slowls withdraws, the vibrations recede. The music becomes too loud.
> [...]
> Dipping into the "underworld", one was cleansed.[22]

20 Ibid., pp. 9–11.

21 See Robert J. C. Young: *Colonial Desire: Hybridity in Theory, Culture and Race*. London: Routledge, 1995.

22 Nazareth: *In a Brown Mantle*, pp. 17–20. For a detailed analysis of this passage see Dan Ojwang: The Pleasures of Knowing: Images of 'Africans' in East African

As Dan Ojwang has argued, the racial hierarchies installed and utilized by British colonialism in East Africa played a major role in generating 'racial' stereotypes and social antagonisms: "[T]he very ordering of colonial society made stereotyping almost inevitable. Placed and constrained in specific economic roles, between which there was little room for mobility, blacks and Indians came to occupy mutually hostile subject positions, the bridging of which required leaps of the imagination."[23]

While such 'leaps of the imagination' undoubtedly occurred in late colonial and early independent East Africa (*In A Brown Mantle* can itself be read as such a 'leap' in the fictional realm), the exoticist imagery in the above passage from Nazareth's novel should serve as a salutary reminder that 'Indian Ocean Imaginaries' come in a wide variety of shapes and functions and can certainly not be restricted to a 'beneficial' vision of trans-regional exchange alone. What is at play in the nightclub scene in *In a Brown Mantle* is arguably an Indian Ocean Imaginary predicated on an Asian fantasy of Africa as a 'wild' continent and a notion of India's 'civilising mission' in Africa that strongly echoes nineteenth and twentieth century European colonial discourses. An intriguing example of this problematic Indian Ocean Imaginary can be found in a passage in M. G. Vassanji's novel *The Book of Secrets*, where one of the protagonists, an Indian teacher recruited to teach in colonial Zanzibar, reflects on the naïve vision of Africa that once inspired him and his colleagues to cross the Indian Ocean:

> We had been told that all three of us had been posted to the government school in the historic town of Tabora, in the interior of the country. We did not know what to expect, none of us having taught at school, let alone in a part of the world about which we had only the faintest notion – and a lot of fantasies culled from the likes of Rider Haggard, Tarzan, and Sanders of Africa. [...] We were sailing to freedom: freedom from an old country with ancient ways, from the tentacles of clinging families with numerous wants and myriads of conventions; freedom even from ourselves grounded in those ancient ways.[24]

Just like the Indian teacher in Vassanji's novel, many of the Goan characters in Nazareth's text struggle to overcome these exoticist fantasies,

Asian Literature. In: *English Studies In Africa: A Journal Of The Humanities* 43,1 (2000), pp. 43–64.

23 Dan Ojwang: *Reading Migration and Culture: The World of East African Indian Literature*. Houndsmills / Basingstoke: Palgrave Macmillan 2013, p. 3.

24 M. G. Vassanji: *The Book of Secrets*. New York: Picador, 1997 [1994], pp. 238–239.

however, and *In a Brown Mantle* is centrally concerned with D'Souza's 'leap of the imagination' that leads to his involvement in the anticolonial struggle. If the Goan-African characters (and particularly the unreliable narrator Deogratius D'Souza) are no immaculate heroes in Nazareth's novel, neither are the black African politicians, many of whom are quick to give up their democratic ideals once they have arrived in positions of power. As the novel progresses, D'Souza finds himself increasingly ostracised by his 'Africanist' opponents in the government who consciously play the race card to drum up political support: his battle against colonial oppression thus quickly turns into a struggle for citizenship in independent Damibia. In a telling scene set in a Club frequented by the new power elite, D'Souza meets a cabinet minister nicknamed "the Cow" who conspires with the military against Kyeyune, Damibia's first African president:

> When I walked in, I saw a group in the corner drinking away and saying "Cheers!" very loudly. It was the Cow and the two leaders of the Army. Hallo! It was an unwritten law that the army, even the leaders, did not visit the Damibia Club and did not fraternize with the ministers. As I was walking to a seat, The Cow called out,
>
> "Hey, Mugoa! When are you going back to your country?"
>
> "When the cat's away, the rats are at play," I mumbled.
>
> "Have you met Colonel Makenju and Brigadier Karogo? Gentlemen, meet Mr. – Mr. Patel–"
>
> "D'Souza," I said, angrily.
>
> […]
>
> "Gentlemen," The Cow said, "D'Souza here has finished his work for Damibia and will soon be returning to his country. When are you leaving?" It was clear that The Cow was drunk because he normally took care to maintain cordial relations with me.
>
> "I was not aware that I was leaving," I said.
>
> "Don't you love your country?" said The Cow.
>
> "I do," I said "Damibia."
>
> "If I were in power," said The Cow, "I would chase all these brown people into the sea." He laughed, as though he was joking, but I was sure he was deadly serious.
>
> "They kept aloof from us until we won our Independence and now that we are the bosses, they are trying to be friendly."
>
> "You know that is certainly not true in my case," I said, stung. "I worked with Kyeyune long before Independence."
>
> "Ah yes, Kyeyune," he said, staring at the ceiling. "A very kind man. A very tolerant man. Too tolerant."[25]

25 Nazareth: *In a Brown Mantle*, pp. 113–114.

While the text here is clearly intent on exposing (and delegitimising) a new type of African racism fanned and instrumentalised by an increasingly reckless political caste bent on maintaining its own political and economic privileges (a racism that prefigures the political discourse prevalent during Idi Amin's reign of terror in non-fictional Uganda), Nazareth's novel consciously steers clear of collectivist ascriptions and refrains from turning the political conflicts in post-independence Damibia into a simple antagonistic tale of 'Africans' vs. 'Asians'. What is at issue is a struggle over the definition of citizenship in which D'Souza (and many of his Goan-Damibian compatriots) eventually lose out and are turned into diasporic subjects once again. Yet the novel argues clearly that on the one hand not all Goan-Damibians are engaged in this struggle for citizenship and that, on the other hand, not all (black) Africans concur with the logics of racial separation and excision that underlie the thought and actions of politicians like "The Cow". When D'Souza finally leaves Damibia, he joins a diasporic network linking Goa to Africa, North America and Europe, but he is also going into exile. His last farewell constitutes a bitter acknowledgement of his contested 'racial' status and reconfirms his claim to African citizenship:

> I had to leave Damibia. I was sickened beyond cure of the political scene. [...]
> My mother was already out of the country, having gone to Goa. My brother was studying in Germany. [...]
> "Goodbye, Mother Africa," I said, as the plane lifted off. "Your bastard son loved you."[26]

Conclusion

Peter Nazareth's tale of former political allies turning into 'intimate enemies' in post-independence East Africa is acutely aware of the complicated legacies of political, social and cultural South Asian / African interactions, and studying *In a Brown Mantle* (as well as other works by Asian-East African writers such as M.G. Vassanji) in more detail arguably offers excellent opportunities to move beyond some of the limitations of contemporary Indian Ocean studies such as Third Worldist naïveté and the dichotomy of 'romantic stereotype' or 'silence' in the representation of Asian-African relations that Isabel

26 Nazareth: *In a Brown Mantle*, p. 150.

Hofmeyr pinpointed in her critique cited in the first part of this essay. Yet the story of South Asian-African conflict in post-independence East Africa is above all a story of Africa's Asian options that were either totally rejected (as in Amin's Uganda) or muted (as in Kenya or Tanzania) in the decades following independence.

If Indian Ocean studies are not only to address the distant past of precolonial contact, the intermediate past of nineteenth and twentieth century interactions in the context of colonialism and the immediate past of post-independence Asian-African conflict, but also to explore the present and possible future of trans-regional African-Asian interactions, they will also have to come to terms with the Indian Ocean imaginaries produced in literature and other media that engage with contemporary East African life-worlds. If and how these imaginaries intersect with those relating to earlier phases of history and whether they will extend beyond the established East African-South Asian nexus to encompass further African-Asian interactions remain challenging questions for future research.

Bibliography

Desai, Gaurav: Oceans Connect: The Indian Ocean and African Identities. In: *Publications of the Modern Language Association of America* 135,3 (2010), pp. 713–720.

Farah, Nuruddin: Of Tamarind & Cosmopolitanism. In: Ntone Edjabe / Edgar Pieterse (eds): *African Cities Reader*. Johannesburg: African Centre for Cities / Chimurenga Magazine 2010, pp. 9–12.

Gupta, Pamila: Introduction. In: Ead. / Isabel Hofmeyr / Michael Pearson (eds): *Eyes Across the Water: Navigating the Indian Ocean*. Pretoria: Unisa 2010, pp. 3–4.

Gurnah, Abdulrazak: *By the Sea*. London: Bloomsbury 2002 [2001].

Hawley John C.: Introduction: Unrecorded Lives. In: John C. Hawley (ed.): *India in Africa, Africa in India: Indian Ocean Cosmopolitanisms*. Bloomington / Indianapolis: Indiana University Press 2008, pp. 1–15.

Hofmeyr, Isabel: Africa as a Fault Line in the Indian Ocean. In: Ead. / Pamila Gupta / Michael Pearson (eds): *Eyes Across the Water: Navigating the Indian Ocean*. Pretoria: Unisa 2010, pp. 99–108.

Hofmeyr, Isabel: Universalizing the Indian Ocean. In: *Publications of the Modern Language Association of America* 125,3 (2010), pp. 721–729.

Kaplan, Robert D.: *Monsoon: The Indian Ocean and the Future of American Power*. New York: Random House 2010.

Nazareth, Peter: *In a Brown Mantle*. Nairobi: East African Literature Bureau 1972.

Ojwang, Dan: The Pleasures of Knowing: Images of 'Africans' in East African Asian Literature. In: *English Studies In Africa: A Journal Of The Humanities* 43,1 (2000), pp. 43–64.

—: *Reading Migration and Culture: The World of East African Indian Literature*. Houndsmills / Basingstoke: Palgrave Macmillan 2013.

Rastogi, Pallavi: *Afrindian Fictions: Diaspora, Race, and National Desire in South Africa*. Columbus: Ohio State University 2008.

Vassanji, M. G.: *The Book of Secrets*. New York: Picador 1997 [1994].

Wawrzinek, Jennifer / J. K. S. Makokha (eds): *Negotiating Afropolitanism: Essays on Borders and Spaces in Contemporary African Literature and Folklore*. Amsterdam / New York: Rodopi 2011.

Wright, Michelle M.: Middle Passage Blackness and its Diasporic Discontents: The Case for a Post-War Epistemology. In: Eve Rosenhaft / Robbie Aitken (eds): *Africa in Europe: Studies in Transnational Practice in the Long Twentieth Century*. Liverpool: Liverpool University Press 2013, pp. 217–233.

Fabulating the Indian Ocean – An Emerging Network of Imaginaries?

Ute Fendler

It could surprise to talk about an "emerging network of imaginaries" in the context of the Indian Ocean as researchers from various disciplines rather underline the centuries of exchange in this very particular contact zone. Sugata Bose sketches the temporal dimension of the region as follows:

> The Indian Ocean has been traversed by a number of distinguished historians with a penchant for describing long- and medium-term movements in trade and culture. Whether they have taken on a whole millennium or just a couple of centuries, most have chosen to concentrate on the pre-modern and early modern periods. What emerges from these studies of the Indian Ocean until the eighteenth century is a picture of a well-integrated interregional arena of economic and cultural interaction and exchange. Particularly important connections of material life, politico-military organization, economic institutions, and social-religious ideology were forged across the ocean during the millennium that stretched from the eighth to the eighteenth century. The modification of theses links in the late eighteenths and early nineteenth centuries critically influenced the nature of the colonial transition in South and South-east Asia and European ascendancy in the Middle East.[1]

Isabel Hofmeyr describes the same setting but brings forward the need to question certain paradigms by integrating cultural and literary perspectives.[2] She therefore suggests as alternative and / or additional

1 Sugata Bose: *A Hundred Horizons. The Indian Ocean in the Age of Global Empire*. Cambridge, MA / London: Harvard University Press 2006, p. 15.

2 Isabel Hofmeyr: The Black Atlantic Meets the Indian Ocean: Forging New Paradigms of Transnationalism for the Global South – Literary and Cultural Perspectives. In: *Social Dynamics* 33,2 (2007), pp. 3–32.

paradigms "people", "diaspora" or "islands" arguing that these reflections will also allow having a new perspective on the literatures of Southern Africa.[3] She brings here together three aspects that we are particularly interested in as the question of how the suspected disruption of the Indian Ocean as a shared regional space in the nineteenth century under the influence of the shifts in a world of the predominance of imperialistic and economic interests might be overcome via the creation of a shared imaginary of the Indian ocean as a common arena in narrative texts. Literature is considered to be a privileged space where limits of time and space can be overcome and new imaginary spaces can be created. But the question here would not be how the Indian Ocean perspective would allow emerging a Southern African literature but the necessary shifts mentioned earlier shall be combined with approaches from Indian Ocean studies. Shanti Moorthy and Ashraf Jamal claim to focus on the ocean as an "area":

> Our view from the ocean completely inverts the land- and continent-based approach of the older area studies, which leaned heavily on modern and postcolonial nation-states as units of homogenous continents; [...]. The uniquely littoral nature of Indian Ocean societies makes engagements between regions inevitable, so there is no fear of a return to the constraints of an area studies discipline that subscribes to patrolled geographic, historical, or cultural boundaries.[4]

The shift of the perspective from the land to the ocean focuses the littoral, the rim, the contact zone per se of land and sea and of various cultural, historical, social, linguistic as well as economic referential systems. The question is if the long lasting tradition of connections and exchange in this area will overcome the cleavages and frontiers created from the eighteenth century onwards and how does the imaginary dynamics of literature contribute to the (re)construction of an Indian Ocean network. Véronique Bragard summarises recurring aspects in literary constructions in her book *Transoceanic Dialogues*[5] on literatures in the Caribbean Sea and Indian Ocean. Oceans and seas merge in the literary imagination as

3 Hofmeyr: The Black Atlantic Meets the Indian Ocean, p. 4.

4 Shanti Moorthy / Ashraf Jamal (eds): *Indian Ocean Studies. Cultural, Social, and Political Perspectives*. New York / London: Routledge 2010, p. 8.

5 Véronique Bragard: *Transoceanic Dialogues: Coolitude in Caribbean and Indian Ocean Literatures*. Frankfurt am Main: Lang 2008.

> […] immensities from which everything proceeds and to which everything returns, a source of the (re)generation of life. Some cultures even ascribe to the ocean the beginning of life itself. Given the essential place of water in life and its link with primordial creation, water imagery plays a prominent role in literature and myth. […] The ambivalence of the oceanic site is associated with death, introspection, but also plenitude, immoderation. […] The sea is also a symbol of unfathomable truth and wisdom. […] This symbolic meaning posits the sea as the collective unconscious, related to emotions, desires and longing for spiritual experiment. Oceans are also fundamentally associated with western colonial history. This image of the ocean as the location of expansionist adventure is currently defied by descendants of slaves or indentured labourers.[6]

The imagery of the sea in literature oscillates between these extremes of wisdom and origin on the one hand and loss of the past and memory on the other. Very often the reflections on the significance of the ocean or the sea are linked to the island as being a space in between land and sea. As Frank Lestringant shows for the imagery of European literature, the narrations of archipelagos and islands illustrate a transitional space between a landmass and the sea, in a certain way a "space in between":

> C'est que l'île revêt une ambigüité constitutive : formée de terre, elle se définit néanmoins par la mer, le lac ou la rivière qui l'entoure. Sa nature est donc essentiellement hybride. […]
> Une étymologie erronée, mais généralement reçue, décomposait le mot latin insula en in- et salo. In salo « dans la mer ».[7]

The islands are therefore the privileged position from where reflections on and narrations about the sea are created. Examples taken from some texts from the Indian Ocean, where the literary construction of the sea or the ocean is a topos in the literatures of islands like Mauritius, represent a corpus that allow dealing with the questions presented before. Texts from Madagascar, although being an island, deal with the sea in a different way, and are to a certain extent closer to literature from the African continent (as e.g. from Mozambique) in their way of focusing on the line between land and sea. In spite of these differences, the various countries mentioned, Mauritius, Madagascar, Mozambique, do belong to the Indian Ocean area, so that the question could be raised how the literary constructions of the sea or the ocean

6 Bragard: *Transoceanic Dialogues*, p. 66.

7 Frank Lestringant: *Le livre des îles: Atlas et récits insulaires, de la Genèse à Jules Verne.* Genève: Les seuils de la modernité 2002, p. 26.

may contribute to the construction of a shared imagery of the Indian Ocean—and beyond with the Caribbean.

1. Narrating the Sea from the Island: The Perspective from Mauritius

Being an island is linked to notions of being lost in space and time, being marginal. It might therefore not be surprising that two myths have been created in this context that link the island to a lost continent—similar to Atlantis. Jules Hermann, a poet and researcher from Reunion Island, dedicated a large part of his work to a kind of historical reconstitution of material that would allow proving that the islands of the Indian Ocean are just the remaining parts of a lost continent "Lémurie".[8] He links this idea to studies of the 'proto-continent' Gondwana of which the islands would be—according to his interpretation—the leftover of this continent. Still, the belonging to a continent—even one that has disappeared—changes the position:

> L'exil colonial change de sens : l'île n'est plus à la périphérie du monde, mais s'identifie avec son centre même, puisqu'elle fut le berceau d'une race humaine dont celle de l'Europe peut-être issue ; de sorte que l'Européen expatrié aux Mascareignes, renouant avec sa véritable origine oubliée, peut se rêver lointain descendant des Lémuriens, autochtone de l'île, et par conséquent son légitime possesseur, en même temps que l'héritier d'une histoire prestigieuse, infiniment plus ancienne que celle du vieux monde, dans une terre rétrospectivement investie du privilège continental.[9]

Later on, this mythological construction of belonging will be found again in the texts of Malcolm Chazal from Mauritius as, for example, in *Petrusmok*.[10] And there are echoes of the construction of this myth in the poem with the title "Géographies d'un exil" by the Mauritian poet Edouard Maunick. The four stanzas are numerated as "place 1,2,3,4" which might refer to the four compass points, a nautical expression. The second stanza though starts with a vague spatial indication of 'somewhere', a piece of Gondwana or a lost part of Lémurie. The fact that the island does not belong to a continent makes it a plaything of the winds and the sea with changing names and therefore

8 Jean-Michel Racault: *Mémoires du Grand Océan. Des relations de voyages aux littératures francophones de l'océan Indien.* Paris: Presses de l'Université Paris-Sorbonne 2007, pp. 23–24.

9 Racault: *Mémoires*, p. 25.

10 Malcolm de Chazal: *Petrusmok.* Paris: Éditions Léo Scheer 2004 [1951].

changing affiliations. Maunick invites the reader to complete the gaps in the narration, to bring together this image of a floating island with the various names gone with the wind and the traveller. The sea is not mentioned explicitly but is rather the context in which the whole narration is embedded:

> Lieu 2
> L'ÎLE était quelque part
> débris de Gondwana
> ou bris de Lémurie
> terre non appartenue
> baptisée sans baptême
> de noms de traversée
> arabo-swahili
> Dina Moraze ou
> Dina Mashriq
> échos emportés par vents d'Est
> depuis près de dix siècles
> comme Cirné ou Cerné
> mais suivront d'autres noces
> préludes incantatoires
> d'un creuset trinitaire
> afroasiaeuro
> de Stella clavisque
> maris indici (sic) :
> rumeur avant-coureuse
> d'une saga insulaire[11]

Maunick also mentions the motto of the republic of Mauritius: "Stella clavisque maris indici" ("Star and key of the Indian ocean") which puts the focus on being lost in the ocean, a small spot that is only part of the ocean, a space where travellers or nomads get into contact more by chance than by premeditation. At the same time, being the star and the key refers to the point of orientation in the vast ocean. The attempt to create a genealogy for the island that would overcome the vision of 'being lost in the ocean' is given up in favour of the idea of a nomadic space crossed by oceanic routes.

When Stephen Muecke launches fabulation as one of the most significant aspects in Indian Ocean studies that brings about "artful politics",

11 Edouard Maunick: Géographies d'un exil. In: Id.: *Brûler à vivre / Brûler à survivre*. Sarcelles: Le Carbet / Maison de l'Outre-Mer 2004, pp. 14–15.

he puts the focus on the "listening" and the "telling" of voices and stories that "project reality":

> Fabulation involves inventing in relation to a problem or a situation that must be remedied, particularly with regard to the situation that of "the people who are missing". [...] So fabulation involves the writer and the people moving towards one another. [...] Fabulation works best hovering between the oral and written, the real or the imaginary, being neither a document nor a fiction, but a fictocritical form of enunciation that oscillates between these poles. This is not the romance of the indeterminate; it is a listening method which "hears" not only the words but the discursive frameworks which are the practice for putting words together as knowledge.[12]

Muecke refers to Deleuze's reflections in his book *Image-Temps*:

> Ce constat d'un peuple qui manque n'est pas un renoncement au cinéma politique, mais au contraire la nouvelle base sur laquelle il se fonde, dès lors, dans le Tiers-Monde et les minorités. Il faut que l'art, particulièrement l'art cinématographique, participe à cette tâche: non pas s'adresser à un peuple supposé, déjà là, mais contribuer à l'invention d'un peuple. Au moment où le maître, le colonisateur proclament « il n'y a jamais eu de peuple ici », le peuple qui manque est un devenir, il s'invente, [...] dans de nouvelles conditions de lutte auxquelles un art nécessairement politique doit contribuer. [...] Il reste à l'auteur la possibilité de se donner des intercesseurs, c'est à dire de prendre des personnages réels et non fictifs, mais en les mettant eux-mêmes en état de 'fictionner' de 'légender' de 'fabuler'. L'auteur fait un pas vers ses personnages, mais les personnages font un pas vers l'auteur: double devenir. La fabulation n'est pas un mythe impersonnel, mais ce n'est pas non plus une fiction personnelle : c'est une parole en acte, un acte de parole par lequel le personnage ne cesse de franchir la frontière qui séparerait son affaire privée de la politique, et produit lui-même des énoncés collectifs.[13]

The poem of Maunick corresponds to this pattern that Muecke claims for the Indian Ocean: it is a "parole en acte", "words transformed into action", an act of enunciation that produces facts via the performative strength of the words. With the need of inventing one's own history, or origins, fabulation becomes an interactive process between the narration and the reader: the poem invites the reader to follow the references that create a map of journeys of discovery that allow the island to take a concrete form, and even to become a point of orientation.

12 Stephen Muecke: Fabulation: Flying Carpets and Artful Politics in the Indian Ocean. In: S. Moorthy / A. Jamal (eds): *Indian Ocean Studies. Cultural, Social, and Political Perspectives*. New York / London: Routledge 2010, pp. 32–34, here p. 33.

13 Gilles Deleuze: *Cinema 2: Image-Temps*. Paris: Minuit 1985, p. 289.

In addition to this issue of belonging, another recurrent question in Indian Ocean literature is the origin of the immigrated population, which can be linked to the idea of the "missing people" ("l'invention d'un peuple") of Deleuze.[14] The fabulation of the lost origins that invite the concerned people to "fictionalise" ("fictionner"), "to legendise" or to "fabulate" are concerned with the Middle Passage as the moment of the rupture, of the loss. Véronique Bragard points out the importance of the parallel between the Middle Passage narratives of the African slaves brought to the Americas and the narratives of the "Dark waters" (Kala pani) in the Indian Ocean especially in Mauritius. The loss of memory with the loss of culture, language and roots due to the long journey in the dark womb of a ship stands for the beginning of a new 'creolised' society that will come into being.[15]

Khal Torabully, another poet from Mauritius, coins the notion of "Coolitude" as a kind of prolongation of the reflections of the authors and philosophers of the "Créolité" and Creolization in the Caribbean context adding the experience of the Coolie. In his collection of poems *Arbres et anabase*, we find the echo of the "kala pani" as well as the link to the Caribbean by appealing to Césaire, Damas, Glissant:

> Que le bois cierge officie au plus sombre de mon cri
> J'abhorre ce terme du bateau fait de bois de couille
>
> Ce cahot cochon quand les vagues s'ouvrent
> S'offrent en découverte de leur propre errance
> Je dis Damas
> Je dis Césaire
> Je dis Glissant
> Je dis Dabydeen
> Au végétal de cette terre où mémoire sent bois-ti-graines
> Il nous faut relire la méprise de noms entendus
> Reprendre le pacte de poésie à l'entrée du saponaire [16]

The repetition of the act of enunciation underlines the importance of the enactment of the word, conjuring the Caribbean models. Starting from this common experience, Torabully sketches in a few lines the quest for an anchorage after the violence of the passage towards forced labour, which leads to an impasse, a dead end:

14 Ibid., pp. 285–286.

15 Bragard: *Transoceanic Dialogues*, p. 75.

16 Khal Torabully: In. Id.: *Arbres et anabase*. Matoury Cedex: Ibis Rouge 2005, p. 39.

> Nous sommes dans le Grand Cul-de-Sac
> A jonction d'océan et de mer
>
> Peu de gens demeurent
> Héritiers de l'absence de lieux
>
> L'arbre localise le souffle libre
> Pour peupler la terre espérée
> Dans les navigations contre l'adversité.[17]

He overcomes the dead end by the symbol of the tree, though activating at the same time the image of roots that link a plant or the human being deeply to a certain spot of earth. The feeling of being lost in the vast ocean, living on the rim or on a floating bit of earth seems to conjure the need of finding a strong attachment to the island. However, this image contradicts the idea of Edouard Glissant expressed in the notion of "poétique de la relation" opposing the unique root with mangrove-like roots that spread on a horizontal level linking various plants with each other. This is surprising as Torabully had established before a parallel between the Caribbean and the Indian ocean via the traumatising experience of the Middle Passage, the shipping of slaves from the African continent to the Americas. In his book on coolitude he underlines the

> concept of the ocean as a nodal moment of migration, a space for deconstruction of identity, yet also one of regeneration, when an aesthetics of migration was created.[18]

The ocean becomes the space of negotiation of memories and identities where islands are the meeting point of the passages that reinforce their being. In Torabully's case the Indian Ocean opens upon parallels drawn with the Caribbean, and also with Africa, which moves towards Glissant's "tout-monde", an aspect that can also be found in the context of Madagascar.

2. Narrating the Sea from the "Continent": The Perspective from Madagascar

Madagascar is a continent-island, 'la grande île', the big island that suffices itself. Having known a very different history of migration and also an autochthonous population, the idea of memory, the question

17 Marina Carter / Khal Torabully: *Coolitude: An Anthology of the Indian Labour Diaspora.* London: Anthem Press 2002, p. 32.
18 Ibid., p. 17.

of origins is put in a different way. According to oral traditions the autochtonous people are the so called Vazimba, most probably of African origin, who are dwarf like and of dark skin, being like ghosts considered as completely different from human beings and in command of healing forces but also evil power. Magali Marson explains the linkage of this figure in traditional folk:

> Mettant à nu une Histoire d'écorchures et de non-dits, dévoilant la mémoire fêlée de l'île, la surnature nous indique une direction particulière, la seule certitude finalement : celle de la mer, « rano masina », salée, sacrée, consacrée et frappée d'interdit donc. La surnature en fait le berceau et la mémoire de l'île ; ne l'oublions pas, les Vazimba ne demeurent pas ici (ils) viennent de dedans la mer.[19]

Oral traditions reveal that the inhabitants of Madagascar are all coming from the sea: the Vazimba as well as the nobles: the origins of the island dweller "hors sol, ex-île […]" ("out of the territory, ex-island").[20] Marson refers to the novel *Nour, 1947* by the Madegassan author Jean-Luc Raharimanana who takes up pieces of a legend, namely that of Ranoro, or Nour, a mermaid who marries a man under the condition that he will never reveal her origin, being a "child-of-the salt" ("enfant-du-sel").[21] The sea is a sacred, a forbidden place, place of the dead and at the same time linked to the horizon which might bring about danger but also new opportunities. In the text of Raharimanana, however, the sea and Nour represent the loss of memory:

> Nous croyions que cette île, si immense qu'elle fût, était la Terre entière. Qu'au-delà, par-delà l'horizon, n'existaient que les morts. Nous avons oublié, ou feint d'ignorer, que nous venons d'ailleurs, d'un ailleurs qui nous avait chassés ou poussés sur les mers à bord de nos boutres chétifs. Inventant des origines célestes,

19 Magali Marson: Les Vazimba : une surnature outil et stratégie de la quête et d'une restauration de soi. In: Valérie Magdelaine-Anrianjafitrimo / Jean Claude Carpanin Marimoutou / Bernard Terramorsi (eds): *Démons et Merveilles, Le Surnaturel dans l'Océan Indien.* Actes du Colloque international, Université de La Réunion, 26–29 Octobre 2004. La Réunion: Océan Éditions 2006, pp. 133–154.

20 Ibid., p. 150.

21 Cf. Valérie Magdelaine-Anrianjafitrimo: Madagascar, 29 mars 1947: Tabatabà ou parole des temps troubles. In: *E-rea,* 8,3 (2011). http://erea.revues.org/1835 (accessed October 21, 2013). "Il existe en effet deux Noro dans la mythologie malgache, la Ranoro qui appartient à la mythologie fluviale des hauts plateaux, la sirène mère mythique, et la Noro d'Ethiopie, noire. En faisant de sa Nour une esclave et fille d'esclave noir, Raharimanana transgresse l'un des grands tabous malgaches : il désacralise le symbole de la noblesse en lui donnant une ascendance servile."

> créant des mythes nouveaux, nous avons effacé notre passé, occulté notre véritable histoire.[22]
>
> Scintillements sur l'océan: les boutres des pêcheurs…
> Scintillements pour ne point perdre le lien…
> Pouvons-nous dire que nous étions toujours attachés à cette terre ? Ne venions-nous pas d'un ailleurs perdu à regret et que nous avions préféré éliminer de nos souvenirs ? Maintenant, nous sommes sur ces rivages. Au milieu de nulle part. Sans contact aucun avec d'autres terres sinon par quelques souffles d'alizé, par quelques odeurs de continent arraché, poussiéreux et volatile. La Terre entière, nous l'avons réduite à cette île si grande ; l'Univers, l'avons peuplé de nos seuls dieux. Et nous nous disons dorénavant maîtres de toute la Terre, souverains universels.[23]

Raharimanana insists on the close relationship between the island and the horizon that opens up to other spaces. The island is conceived as being the universe as there is no link, no contact to other places. In a first step the sea, in this case, was the space of migration, the linking element with an origin, but later became the separating element. He also underlines the rim, the "rivages", the line between land and sea that is a space of transition between natural elements but also between the past and the present, the mythical and the factual. He trespasses this line himself by establishing a link to the African continent, the island being torn apart, the only link given is the wind. And as the wind is often seen as the element that carries stories, the wind is also at the root of the stories about the past, the ancestors, of the "other" beyond the horizon.

The narrator laments that his stories are like rumours that will end on the sandy beaches. As if the sea and the wind would bring these stories ashore, but without any coherence, more like stranded parts of a lost unit:

> Aujourd'hui, mes rumeurs échouent lamentablement sur ces sables. […] Nous n'avons pas su vivre avec l'autre. Ou, simplement, l'autre n'a jamais cru que nous pouvions vivre avec lui. Notre monde a disparu et nous nous dirigeons vers le sien. […] L'irruption de l'autre a faussé le cours de notre histoire et bousculé- que dis-je ? détruit, oui – notre vision du monde.[24]

The consequence will inevitably be the wish to leave the island, this closed universe in order to discover the other. As the horizon stood for

22 Jean-Luc Raharimanana: *Nour, 1947.* Paris: Le Serpent à Plumes 2003, p. 26.

23 Ibid., pp. 43–44.

24 Ibid., p. 27.

the space of the dead, the narrator has to overcome this limit hindering the population to set out for new endeavours, new encounters:

> Nous voudrions partir, voir d'autres mondes. Savoir ce qu'il y a au-delà de ces mers quand il ne serait plus possible de dire que là reposent tout simplement les morts. L'horizon ne serait plus l'espace des songes et des dieux. Il serait juste celui de l'autre.[25]

> Ainsi s'est toujours comportée cette île en marge: se perdre dans les eaux plutôt que dans la servitude, choisir le suicide plutôt que le martyre. Partir. Rejoindre les terres mythiques alliées au ciel.[26]

Raharimanana conceives of the sea as a space of transition but also as a space of disconnection. The will to invent myths of origin that would place the island in the centre of the world—like in the case of Réunion and Mauritius—discards the arrival from elsewhere, veiling a history of migration with unknown motives. Coming from a family of storytellers and a family where various ethnic and cultural descendants come together, Raharimanana retells the history of the island by mingling various oral sources. His main concern is the memory of the Madegassan people not as a multiplicity of ethnic groups but as the inhabitants of the island. Valérie Magdelaine-Andrianjafitrimo remarks that this fusion of the various origins is though problematic in the context of Madagascar:

> Du coup, quelle est la lisibilité du texte d'un auteur qui, par ailleurs, est peu et mal reçu à Madagascar ? Il fait même, pour ce roman, l'objet d'une « forclusion inévitable » en raison de sa façon d'appuyer sur cet événement traumatique et de le mettre en relation avec le grand tabou de l'esclavage et des divisions ethniques. L'auteur, rappelons-le, cherchait une « forme pour contourner la surdité occidentale et les préjugés sur soi. [Une] forme afin de coller au plus près du sens et dire à notre manière le récit de nos vies », mais a-t-il réussi à lutter contre l' 'immémoire' qu'il redoute tant ?[27]

In his essay *Arbre anthropophage*, Raharimanana evokes the ancestors of his own family which in a certain way could be representative of the island:

> Car je suis le Bâtard. Fils d'Orient et d'Occident. Fils d'Afrique et d'Arabie. Je suis l'eau mêlée des lacs jumeaux que l'on invoque en secret. [...] nous venons de l'horizon. Nous venons des cieux alliés à la terre. Cette île nous appartient...

25 Ibid., p. 38.
26 Ibid., p. 39.
27 Magdelaine-Anrianjafitrimo: Madagascar, 29 mars 1947.

> Car le mythe n'est pas des temps reculés mais d'Aujourd'hui. Rêves de pureté et de grandeur. Pour un Présent de mensonge et d'imposture. Pour voiler ce qui hier a fait de nous des fugitifs des océans. Pour voiler ce qui hier nous a poussés sur ces rivages.
> Les conquérants toujours reviennent pour chanter leurs découvertes. Où sont nos terres d'origine pour que nous entamions nos chants ?
> L'Asie ! Ah ! l'Asie. Nusantara. Gondwana…
> Et nous, qui ne sommes qu'ici. Sur cette île de perdition où l'on se déchire. A nous raconter nos exploits passés rêvés. A nous revendiquer des civilisations fabuleuses, maîtres de tout le Pacifique et de l'océan Indien. L'aire culturelle la plus vaste et la plus ancienne du monde.
> Île![28]

Raharimanana is one of the rare authors of Madagascar who questions the myth of origins and focuses on the encounter of various groups and cultures coming from the ocean. The construction of a mythical island as the center of the world might hide a shameful history. There is no answer to his questions, but it allows him to relativize the importance of the noble island, putting it into relation with the Indian Ocean as a small part of a larger area.

In this process of linking Madagascar with other parts of the Indian Ocean, he is also one of the rare authors who evoke Africa as a constitutive part of the cultural and ethnic history of the island. Usually, Madagascar will establish linkages rather with Asia or Arabia, both of which—for not having been enslaved—are considered as being nobler than Africa. In general, the memory of African slaves and the presence of their descendants are being—often voluntarily—ignored, as the system of casts in Madagascar would exclude the human beings that find themselves at the lowest level of the recognition in society. Raharimanana tries to integrate Black persons into his works, as e.g. Nour in the novel *Nour, 1947* is no longer a noble person, but he gives this name to the daughter of a black slave.

His willing to place Madagascar in a larger African context and his preoccupation with the signification of the island as part of a larger region linked by the ocean pushes him to tell stories about other countries that dispose of parallels in their history: e.g. Haiti reoccurs in Raharimanana's texts. In his collection of short stories *Rêves sous le*

28 Jean-Luc Raharimanana: *Arbre Anthropophage*. Paris: Éditions Joelle Losfeld 2004, pp. 60–61.

linceul,[29] the corpse of a Haitian is washed ashore. The story establishes links between the Haitian boat people and the Madagascans Raharimanana describes: in both cases, the horizon is a promise of escape from a confined space.

In *Les cauchemars du gecko*[30] Raharimanana also dedicates a chapter to Haiti and the legend of the jar of gold that the master buried together with the slave for not having any witness. Other parts focus on Rwanda or Congo, developing the African linkage of Madagascar. Here again the process of fabulation can be recognised, as some "real characters" reach out to meet the reader so that a common imaginary can be built. And just to remind us of how Muecke put it:

> This is not the romance of the indeterminate; it is a listening method which 'hears' not only the words but the discursive frameworks which are the practice for putting words together as knowledge.[31]

The narrations of Raharimanana seem to be listening ("à l'écoute") so that he could bring together dispersed elements that once assembled will allow constructing knowledge and memories.

3. Narrating the Sea from the African Continent: The Perspective from Mozambique

If it comes to the African Continent, artists from Angola for example, like António Ole would evoke the sea as a common space of memory of the trade. His photos of the "the door of no return" at the fort on the island of Gorée thematises the openness of the horizon put next to a bricked door that reminds the viewer of the changing significance of the open door and the view on the horizon brings up the experience of the Black Atlantic.[32]

In Mozambique, though, a country with a long coast on the Indian Ocean that also has a history of slavery, the remembrance of the deportation is less present, which might also explain why no Mozambican artist has dealt the subject as António Ole did. The house of the

29 Jean-Luc Raharimanana: *Rêves sur le linceul.* Paris: Le Serpent à Plumes 1998.

30 Jean-Luc Raharimanana: *Les cauchemars du gecko.* La Roque d'Anthéron: Vents d'Ailleurs 2011.

31 Muecke: Fabulation, p. 33.

32 Ulf Vierke / Johannes Hossfeld (eds): *António Ole. Hidden Pages.* Catalogue for the same-titled exhibition of the Iwalewa-Haus Bayreuth. Frankfurt am Main: Hammer 2009, p. 81.

slaves in the city of Inhambane for example is turned to the arm of the river Malumba that leads to the open ocean, almost forgotten. Like this, it hides itself from the open view on the ocean and is turned towards the continent from where slaves had been brought to the coast. But it does not deal with the ocean or the passage.

In literature, we find texts that bring up the ocean as the space of escape from the war-ridden country like in Mia Couto's *Terra sonâmbula*.[33] Mia Couto is the Mozambican author with the greatest international reputation and received the prestigious Portuguese Camões prize for literature in 2013. In his large oeuvre of short stories and novels, he often deals with issues of identity and memory. In this sense, he is known for bringing in strong Indian characters who illustrate the close linkage of Mozambique with India across the Ocean and across epochs.[34]

Nevertheless, Couto also uses the image of a shipwreck where individuals are waiting for lost persons, lost memories, a state in between, not really offshore and not really getting to the coast. In his novel *O último voo do flamingo*,[35] this imagery is again present: the father of the protagonist who returns to his home village faces the ocean reflecting on time and hidden powers that determine the destiny of men. The ocean is a space that represents memory, but also the access to parallel, often mythic worlds.

An author like Ungulani Ba Ka Khosa[36] uses the ocean as a space that is only approached by fishermen who do have a specific knowledge of how to deal with the dangers of the ocean. It is another world—similar to the presence of the sea in the Madegassan texts—a world that is rather one of mythical creatures and significations. In his collection

33 Mia Couto: *Terra Sonâmbula*. Lisboa: Caminho. Outras margens 2008 [1992].

34 See e.g. David Brookshaw: Indianos e o índico: o pós-colonialismo transoceânico e internacional em *O Outro Pé da Sereia*, de Mia Couto. In: Margarida Calafate Ribeiro / Maria Paula Meneses (eds): *Moçambique. Das palavras escritas*. Porto: Edições Afrontamento 2008, pp. 129–139.

35 Mia Couto: *O último voo do flamingo*. Maputo: Nadjira 2010 [2000].

36 Ba Ka Khosa is the author of *Ualalapi* (1987), a book that retells the story of the leader Ngungunhane who affronts the European invaders in the end of nineteenth century. The book also speaks of the violence and brutality also due to the acts of Ngungunhane. See e.g. Niyi Afolabi: *Emerging perspectives on Ungulani Ba Ka Khosa. Prophet, Trickster, and Provocateur*. New York: Africa World Press, 2010.

of short stories *Orgia dos Loucos*[37] the second story entitled "A Praga" confronts the reader with the story of a woman who gives birth to her child in a boat:

> [...] a criança que devia nascer em terra firme como toda a criança do mundo, pois nunca se vira por estas terras tsongas uma mulher parir no meio das águas, nessas águas onde os espíritos malignos emergem e os homens de bom porte a elas se atiram com intuito único de aprenderem o segredo das águas, da terra e dos céus durante anos e anos, [...].[38]

The—even involuntary—disregard of the sea brings about a malediction that they will all die of. Like in the case of Raharimanana, the sea is the space of the dead, of spirits that limit mankind to respect the secrets of the foreign element, the sea. The disrespect will cause ruin, suffering and finally death of the members of the family. Without having read each other, their texts build a similar imaginary about the threatening sea.

However, the positive symbolic charging of the island or the sea in the Mozambican context is also possible as Carmen Secco argues in her article on Mozambican poets using the Island of Mozambique as a motif[39] that the island as being a space between the continent and the ocean opens to the horizon and therefore to new opportunities, as a mode of narration that allows a certain kind of escapism facing the repercussions of the long civil war in Mozambique. The island allows projection of wishes on this space in between, away from the continent, though still belonging to it.

The poet Eduardo White goes one step further. In his long poem "Janela para oriente"[40] the narrator is sitting next to his window that gives on to the ocean and stirs his imagination: he links his position,

37 Ungulani Ba Ka Khosa: *Orgia dos Loucos*. Maputo: Asociação dos Escritores 1990.

38 Ungulani Ba Ka Khosa: A Praga. In: Id.: *Orgia dos Loucos*, pp. 19–32, here p. 21. "[...] the child that should be born on firm ground just like every child in the world, for never had anyone in these tsonga lands ever seen a woman bear child in water, in those waters where evil spirits emerge and well built men jump at them with the intention of learning the secret of the water, the earth and the heavens for years on end [...]."

39 Carmen Lucia Tindó Ribeiro Secco: Indicos cantares: o imaginário das ilhas de Mocambique pelas vozes dos poetas. In: Mar Garcia / Felicity Hand / Nazir Can (eds): *Indicities / Indices / Indícios. Hybridations problématiques dans les littératures de l'Océan Indien*. Ille-sur-Têt: Éditions K'A 2010, pp. 165–176.

40 Eduardo White: Janela para oriente. In: Id.: *Antologia Poética. Nudos*. Maputo: Alcance 2010 [1999], pp. 129–146.

the local to various places across the sea, behind the horizon. In this way, he manages to create a new universe across the ocean in a certain way recalling ancient paths across the Indian Ocean. He overcomes the cleavage between the continent and the ocean to reach out to Arabia, the Middle and the Far East. White reconnects with an old imaginary of a shared space where people used to migrate in all directions and exchange goods and ideas, which is used as an opening for Mozambique that suffers from a marginalised position on the African continent.

4. Concluding Remarks

Combining approaches from ocean studies with Indian Ocean studies has the advantage that various shifts of foci are possible and will reveal the complexity of the area. While the focus on the ocean has offered the perspective on the relationship between sea and land and how the respective population will determine the meaning in the lived environment, this approach shows that there are parallels between the three countries that served as a first sample. Literature contributes to the fabulation of a shared space with a wide range of facets that include parallels as well as oppositions. But the ongoing building of a network of narrative lines that link the various layers of meaning contribute to the merging of an idea of what the Indian Ocean as a shared experience could mean. The most interesting point might be that after the interruption of the continuous flow of a common imaginary till the nineteenth century, the narrative texts of the twentieth and twenty-first centuries pick up elements of the shared memory and relate them to new realities. It seems that the symbolic value of the ocean as an open space for the imagination of future encounters becomes more important. Françoise Lionnet suggested thinking together creolization and cosmopolitanism in order to overcome the cleavages that these notions capture:

> […] we might begin to think of creolisation as the cosmopolitanism of the subaltern, and cosmopolitanism as the creolisation of the elites.[41]

41 Françoise Lionnet: Cosmoplitan or Creole Lives? Globalized Oceans, Insular Identities. In: Id.: *The Known and the Uncertain. Creole Cosmopolitics of the Indian Ocean*. Paris: L'Atelier d'Ecriture (Maurice) 2012, pp. 58–124, here p. 65.

The cosmopolitanism especially of harbour cities in early centuries is one of the characteristics of the Indian Ocean area widely accepted. In modern times, one would though rather use the notion of creolization to capture the specificity of the ongoing process of the encounter of various cultures and peoples. The new forging of shared imaginaries in the Indian Ocean where this area becomes again an area of cultural encounter and the sea rather a relating element than a barrier can be thought together with the notion of "Creole Cosmopolitics" that Lionnet suggests. The fabulation process not only fills the gap but it also carries on the building anew of a shared imaginary across linguistic and historical barriers.

Bibliography

Afolabi, Niyi: *Emerging Perspectives on Ungulani Ba Ka Khosa. Prophet, Trickster, and Provocateur.* New York: Africa World Press 2010.

Ba Ka Khosa, Ungulani: *Orgia dos Loucos.* Maputo: Asociação dos Escritores 1990.

Bose, Sugata: *A Hundred Horizons. The Indian Ocean in the Age of Global Empire.* Cambridge, MA / London, England: Harvard University Press 2006.

Bragard, Véronique: *Transoceanic Dialogues: Coolitude in Caribbean and Indian Ocean Literatures.* Frankfurt am Main: Lang 2008.

Brookshaw, David: Indianos e o índico: o pós-colonialismo transoceânico e internacional em *O Outro Pé da Sereia*, de Mia Couto. In: Margarida Calafate Ribeiro / Maria Paula Meneses (eds): *Moçambique. Das palavras escritas.* Porto: Edições Afrontamento 2008, pp. 129–139.

Brugioni, Elena: *Mia Couto. Representação, História(s) e Pós-colonialidade.* Edições humus 2012.

Carter, Marina / Khal Torabully: *Coolitude: An Anthology of the Indian Labour Diaspora.* London: Anthem Press 2002.

De Chazal, Malcolm: *Petrusmok.* Paris: Éditions Léo Scheer 2004 [1951].

Couto, Mia: *Terra Sonâmbula.* Lisboa: Caminho. Outras margens 2008 [1992].

—: *O último voo do flamingo.* Maputo: Nadjira 2010 [2000].

Deleuze, Gilles: *Cinema 2: Image-Temps.* Paris: Minuit 1985.

Fendler, Ute: L'île-archipel, le Madagascar polyphonique de Raharimanana : nœud de communication et récepteur mondial. In: Ottmar Ette / Gesine Müller (eds): *Worldwide. Archipels de la mondialisation.* Frankfurt am Main: Vervuert 2012, pp. 93–112.

—: Une écriture relationnelle: violence et histoire coloniale dans l'oeuvre de Raharimanana. In: *Francophonies* 23 (2013), pp. 79–96.

Hofmeyr, Isabel: The Black Atlantic Meets the Indian Ocean: Forging New Paradigms of Transnationalism for the Global South – Literary and Cultural Perspectives. In: *Social Dynamics* 33,2 (2007), pp. 3–32.

Kumari R. Issur / Vinesh Y. Hookoomsing (eds): L'Océan Indien dans les littératures francophones. Pays réels, pays rêves, pays révèles. Paris: Karthala / Réduit: Presses de l'Universitaires de Maurice 2001.

Lestringant, Frank: *Le livre des îles: Atlas et récits insulaires, de la Genèse à Jules Verne.* Genève: Les seuils de la modernité 2002.

Lionnet, Françoise: Cosmoplitan or Creole Lives? Globalized Oceans, Insular Identities. In: Id.: *The Known and the Uncertain. Creole Cosmopolitics of the Indian Ocean.* Paris: L'Atelier d'Ecriture (Maurice) 2012, pp. 58–124.

Magdelaine-Anrianjafitrimo, Valérie: Madagascar, 29 mars 1947: Tabataba ou parole des temps troubles. In: *E-rea* 8,3 (2011). http://erea.revues.org/1835 (accessed October 21, 2013).

Marson, Magali: Les Vazimba : une surnature outil et stratégie de la quête et d'une restauration de soi. In: Valérie Magdelaine-Anrianjafitrimo / Jean Claude Carpanin Marimoutou / Bernard Terramorsi (eds): *Démons et Merveilles, Le Surnaturel dans l'Océan Indien.* Actes du Colloque international, Université de La Réunion, 26–29 Octobre 2004. La Réunion: Océan Éditions 2006, pp. 133–154.

Maunick, Edouard: *Brûler à vivre / Brûler à survivre.* Sarcelles: Le Carbet / Maison de l'Outre-Mer 2004.

Moorthy, Shanti / Ashraf Jamal (eds): *Indian Ocean Studies. Cultural, Social, and Political Perspectives.* New York / London: Routledge 2010.

Muecke, Stephen: Fabulation: Flying Carpets and Artful Politics in the Indian Ocean. In: Shanti Moorthy / Ashraf Jamal (eds): *Indian Ocean Studies. Cultural, Social, and Political Perspectives.* New York / London: Routledge 2010, pp. 32–44.

Racault, Jean-Michel: *Mémoires du Grand Océan. Des relations de voyages aux littératures francophones de l'océan Indien.* Paris: Presses de l'Université Paris-Sorbonne 2007.

Raharimanana, Jean-Luc: *Rêves sous le linceul.* Paris: Le Serpent à Plumes 1998.

—: *Nour, 1947.* Paris: Le Serpent à Plumes 2003.

—: *Arbre Anthropophage.* Paris: Éditions Joelle Losfeld 2004.

—: *Les cauchemars du gecko.* La Roque d'Anthéron: Vents d'Ailleurs 2011.

Tindó Ribeiro Secco, Carmen Lucia: Indicos cantares: o imaginário das ilhas de Moçambique pelas vozes dos poetas. In: Mar Garcia / Felicity Hand / Nazir Can (eds): *Indicities / Indices / Indícios. Hybridations problématiques dans les littératures de l'Océan Indien.* Ille-sur-Têt: Éditions K'A 2010, pp. 165–176.

Torabully, Khal: *Arbres et anabase.* Matoury Cedex: Ibis Rouge 2005.

Vierke, Ulf / Johannes Hossfeld (eds): *António Ole. Hidden Pages.* Catalogue for the same-titled exhibition of the Iwalewa-Haus Bayreuth. Frankfurt am Main: Hammer 2009.

White, Eduardo: *Antologia Poética. Nudos.* Maputo: Alcance 2010 [1999].

“Oceans of Pain”

The Sea as Memory and Metaphor in Angolan Poetry and Rap Music

Ana Sobral

Introduction

Considering that Angola was one of the central posts of transatlantic slave trade from the sixteenth to the nineteenth centuries, it is hardly surprising that the sea has featured extensively in the nation’s collective imaginary and has in fact become one of its primary tropes of cultural expression up until the current day. The endurance of the transatlantic slave trade in Angolan memory is not only attributable to the traumatic experience of slavery, but is also connected to its value as a means of articulating a national identity. In effect, metaphors relating to the sea and the slave trade have often been used as a political tool in postcolonial Angola’s turbulent short history. From the struggle for independence from Portuguese colonialism in the 1950s and 60s, to the civil war between 1975 and 2002, to the social tensions of the twenty-first century, images of the sea have served to construct as well as to transform, adapt and ultimately sustain the national identity’s viability.

The following analysis attempts to trace the changes—and highlight the continuities—in the cultural perception of the sea, focusing mainly on poetry from the 1950s to our days. The choice of Angolan poetry as a case study seems particularly fruitful given the favoured status of this form of literary expression since the early manifestations of a national identity in the nineteenth century, and especially during the years of anti-colonial struggle. To this day, many of the country’s cultural icons are poets, starting with the first president of independent Angola, Agostinho Neto, and including well-known names such as Viriato da Cruz, Alda Lara, António Jacinto and Manuel Rui. Indeed, the prominence of poetry as a means of articulating collective grievances and expectations has found continuity in rap music, which currently enjoys great popularity and has received increasing international press in recent years due to its critical messages and civic engagement.[1]

By focusing on the history of the Atlantic Ocean as a means of transporting slaves from the Angolan coast to the Americas, the poets highlight the negative impact of colonialism on the national territory, presenting the transatlantic slave trade as the pivotal experience of destruction—of both identity and memory—that has had repercussions still felt today.[2] On the other hand, the ocean can also be seen as the connecting point between African Angolans and their descendants who were scattered by the slave trade all over the Americas and Europe. The same space that features as a source of loss is thus turned into a powerful means of transnational alliance, appealing to ethnic as well as cultural ties between Africans.

These images become especially interesting when regarded against the historical and political context in which the poems were composed. By comparing poems by famous authors of the years of anti-colonial struggle with more contemporary texts by poets / rappers of the current young generation of Angolans, we can allow them to enter into a dialogue with each other, highlighting the role of the Atlantic as a source of Angolan identity, which itself has undergone significant changes in the past 50 years. This approach will reveal also inter-generational tensions that can be traced back to the country's recent history of conflict and violence.

1 See for example Angolan Rap: When Money Stops Talking. In: *The Economist*, March 31, 2012. http://www.economist.com/node/21551547 (accessed July 1, 2013); Lydia Polgreen: Change Unlikely From Angolan Election, but Discontent Simmers. In: *New York Times*, August 31, 2012, p. 11; Laura Pawson: Angola is Stirred by the Spirit of Revolution. In: *The Guardian*, March 8, 2011. http://www.guardian.co.uk/commentisfree/2011/mar/08/angola-spirit-revolution (accessed July 1, 2013).

2 An analysis of the slave trade as a process of "dismemberment" of the African identity and memory was famously advanced by the eminent Kenyan writer and scholar Ngugi Wa Thiong'o. His arguments find an exemplary illustration in the Angolan case and can in fact be applied to the discussion of most poems presented here. See Ngugi Wa Thiong'o: *Something Torn and New: An African Renaissance*. New York: Basic Books 2009, pp. 1–29.

Angola and the Atlantic: a Brief Historical Overview

The transatlantic slave trade permeates the Angolan experience in manifold ways. According to recent estimates, the majority of slaves transported to the Americas—about 39 percent—departed from West Central Africa, which had Luanda, the capital of Angola, as its main port.[3] The lion's share of slaves landed in Brazil, which by 1800 had "a slave population of nearly a million,... [and] the single largest concentration of slaves with the widest range of jobs, including sugar and coffee cultivation, mining, and work in urban industries".[4] While this does not imply that all the slaves departing from Luanda actually originated from the territory that would later make out Angola, it marks the city as a place that would become inseparably associated with the experience of slavery.

Perhaps one of the most striking and famous testimonies of the early period of enslavement of Africans by European colonial powers comes from Affonso I, fifth ruler of the Kingdom of Congo, which comprised the North of Angola, the Republic of the Congo, the Democratic Republic of the Congo and part of Gabon. Following in the footsteps of his father, Affonso I had converted to Christianity under the influence of Portuguese traders and missionaries. Considering himself a brother of the king of Portugal, he wrote a number of letters in Portuguese that have become some of the most valuable testaments of the 'native' point of view from a period that has been chiefly registered and conveyed with European pens. In a letter dating from 1526, Affonso I touched upon the problem of slavery, which disturbed the order of his kingdom:

> Each day the traders are kidnapping our people – children of this country, sons of our nobles and vassals, even people of our own family... This corruption and depravity are so widespread that our land is entirely depopulated... We need in this kingdom only priests and schoolteachers, and no merchandise, unless it is wine and flour for Mass [...] It is our wish that this kingdom not be a place for the trade or transport of slaves.[5]

3 As a comparison, 20 percent of the slaves were exported from the Bight of Benin, 13 percent from the Bight of Biafra, 10 percent from the Gold Coast, 7 percent from Senegambia, 5 percent from Southeast Africa. See Lisa A. Lindsay: *Captives as Commodities: The Transatlantic Slave Trade.* Upper Saddle River, NJ: Pearson / Prentice Hall 2008, pp. 7–9.

4 Ibid., 7.

5 Quoted in Adam Hochschild: *King Leopold's Ghost: A Story of Greed, Terror and Heroism*

The dramatic tone of this letter highlights the nature of commercial relationships between Portugal and what would later be Angola, which had been initiated through Diogo Cão's expedition up the Congo River in the late fifteenth century. The chief interest of the Portuguese was the slave trade. Accordingly, the Atlantic Ocean became the focal point of contact between merchants and locals: from the sixteenth to the nineteenth centuries, the Portuguese rule over the territory consisted primarily of the capital Luanda, (founded by the explorer Paulo Dias de Novais in 1575), its hinterland and a number of trading posts along the Angolan Atlantic coast. The slave trade was supported by local collaborators especially in Luanda, who formed a strong Creole society based on racial miscegenation and cultural assimilation. These Creoles further established a strong link with the Brazilian bourgeoisie, whom they provided with slaves for the plantations from the mid sixteenth century onwards.[6]

Luanda's Creole elite would come to occupy a pivotal position in the creation of a national identity that is steeped in contradiction. Essentially the rise of what could be regarded as a national consciousness was set off by the abolition of the transatlantic slave trade in 1836: this plunged the country into an economic crisis and caused a serious decline in the power and prestige of the Creole society, leading to growing dissatisfaction with the metropolitan power.[7] Only a few years later, in 1845, the printing press arrived in Angola, stimulating the production of a high number of periodicals in the capital and allowing the discontent of the educated elite to find an expression.[8]

in Colonial Africa. New York: Houghton Mifflin Company 1998, p. 13. On the dramatic impact of the early Portuguese slave trade on the Kingdom of Kongo see also Gerald J. Bender: *Angola Under the Portuguese: The Myth and the Reality*. Trenton, NJ: Africa World Press 2004, pp. 12–18.

6 See David Birmingham: *Empire in Africa: Angola and Its Neighbours*. Athens, OH: Ohio University Press 2006, p. 23; David Birmigham: *Portugal and Africa*. Athens, OH: Ohio University Press 1999, pp. 56, 92; Ana Mafalda Leite: Angola. In: Ead. / Patrick Chabal / Moema Parente Angel / David Brookshaw / Caroline Shaw (eds): *The Postcolonial Literature of Lusophone Africa*. London: Hurst & Company 1996, p. 103.

7 See Leite: Angola, p. 105. Birmingham: *Portugal and Africa*, p. 94.

8 Jacopo Corrado also argues that the Liberal Revolution in Portugal in 1820 and the declaration of independence of Brazil in 1822 had a significant influence on Angola's intellectual Creoles, as the arrival of political refugees from Portugal and Brazil spread liberal ideals in the West African colony. See Jacopo Corrado: The Rise of a New Consciousness: Early Euro-African Voices of Dissent in Colonial Angola. In: *e-JPH* 5,2

By 1849, what has been considered the first work of Angolan literature in Portuguese was published in the country—significantly, this was a volume of poetry by the 'nativist' José da Silva Maia Ferreira, a Luanda born descendant of Portuguese traders (involved in slave trafficking). Although he had spent his formative years in Brazil, Maia Ferreira returned to Angola as an adult. Influenced and inspired by other Creole intellectuals who called themselves "filhos da terra" (sons of the country), his poetry sought to celebrate the specific African traits of his experience, including descriptions of the Angolan landscape.[9] Equally influential was Joaquim Dias Cordeiro da Mata, who labelled his volume of poetry *Delírios* (1888) as "patriotic literature", and wrote a Kimbundu-Portuguese dictionary. Like Maia Ferreira, he promoted the Angolan intellectuals' engagement with African history and national traditions.[10]

These examples illustrate the ambiguous role played by the Atlantic Ocean, and more specifically by the transatlantic slave trade in Angola: the seeds of a national identity and anti-colonial resistance were planted by descendants of slave traders or collaborators, who were themselves influenced by the Brazilian intellectual bourgeoisie, whose wealth had been directly or indirectly acquired by participation in the enslavement of Africans to work the Brazilian plantations. The development of a more critical perspective on colonialism was initiated by the failure of the metropolitan power to continue catering to the needs and priorities of the colony's urban elite. Their resentment as well as their sense of national unity found expression in the printed word, which, as Benedict Anderson has suggested, constituted one of the pillars of the development of the nation as an "imagined community".[11] We can see here how, in the case of Angola, poetry played an important role in projecting a particular image of the nation from its inception.

Angola's autonomous periodicals would eventually be stifled by colonial censorship when the Portuguese fascist José António Oliveira de

(Winter 2007). http://www.brown.edu/Departments/Portuguese_Brazilian_Studies/ejph/html/issue10/html/jcorrado_main.html (accessed July 1, 2013).

9 See Jacopo Corrado: *The Creole Elite and the Rise of Angolan Protonationalism: 1870–1920.* New York: Cambria Press 2008, pp. 161–165.

10 See ibid., pp. 179–190; Leite: Angola, p. 107.

11 See Benedict Anderson: *Imagined Communities: Reflections on the Origin and Spread of Nationalism.* London: Verso 1991, pp. 32–36.

Salazar rose to power in 1932. The new regime established a colonial dynamic based on the suppression of the Creole elite. Sending in settlers and professionals from the metropolis strengthened the Portuguese presence in the colony. Previously the site of departure of African slaves, the Atlantic Ocean now became the point of arrival for a new wave of white Europeans who, rather than cooperating with the local elite, often in fact took over its position in administrative posts.[12] By the 1950s, however, Angola experienced a revival of African and Creole cultures as a clear manifestation of discontent with the colonial rule. Most notable examples include the realm of popular music, with famous Luanda bands such as Ngola Ritmos performing songs in Kimbundu with politically charged lyrics, and the publication of the poetry review *Mensagem*, which gathered many of the poets that would later become emblematic for the independent nation. The poems within this review articulated a strong anti-colonialist sentiment and once again emphasized the African perspective and experiences of the authors. Influenced by Brazilian modernism, the Angolan poets sought in fact to develop a distinct "national literature".[13]
By the 1960s a Portuguese issue of *Mensagem* was being published in Lisbon, the heart of the metropolis, where many students from Angola and other Portuguese colonies in Africa had gathered, fostering anti-colonial sentiment and creating the ideological backbone of resistance movements back home.[14] Once again, the Atlantic Ocean emerged as a connecting element, bringing together disaffected members of the colonies' urban elites—and, in the case of Angola, providing also one of the most powerful literary tropes, to which we will now turn in greater detail.

12 See Birmingham: *Empire in Africa*, p. 78.

13 See Leite: Angola, pp. 142–143; Maria Nazareth Soares Fonseca / Terezinha Taborda Moreira: Panorama das Literaturas Africanas de Língua Portuguesa. *Cadernos CESPUC de Pesquisa, Série Ensaios, n. 16: Literaturas Africanas de Língua Portuguesa*. Belo Horizonte: Set. 2007. http://www.pucminas.br/raca_cor_etnia/ppt/texto_terezinha.pdf (accessed July 1, 2013).

14 A number of poets and intellectuals from Portugal's African colonies who gathered in Lisbon would in fact later join the anti-colonial struggle as fighters. Bender argues that the desire for independence was mainly motivated by frustration, as the colonial regime proved unable to heed to the colonies' appeals for political negotiation: "When their poems, petitions, manifestos, and peaceful protests proved to be ineffective in achieving even minuscule reforms, they turned their talents and energies towards the goal of total national independence." Bender: *Angola under the Portuguese*, p. 154.

Poetic Representations of the Sea: an Analysis in Three Movements

Most surveys of Angolan literature in general, and poetry specifically divide its production into different phases, although the chronological categorization varies slightly. Overall, critics agree that the breaches in terms of theme, style and even aims closely follow the turmoil of historical changes, as the country plunged into a colonial war (from 1961 to 1974), followed by a protracted civil war (from 1975 to 2002) that was in itself strongly influenced by the Cold War's conflict of geopolitical interests. As Laura Cavalcante Padilha has put it, "there is a cyclical element in the game of war and peace in Angola, and the poetic expression becomes an accessory to that unyielding game, representing its diverse faces".[15]

I have opted here for a division into three phases that roughly correspond to the major turning points of Angola's history over the past five decades. By looking at representations of the sea in poems of these different periods we can in fact reconstruct that history, or rather how particular events were processed in the national imaginary. The sea becomes a metaphor for the experiences, hopes and anxieties of the poets, who might ideally be seen—and have often regarded themselves—as spokespersons for the Angolan people.[16]

15 Laura Cavalcante Padilha: Guerra, poesia, estilhaç[ament]os - um olhar para Angola. *Mulemba* 1 (2009), pp. 1–14, here p. 3 (my translation). Padilha herself mainly distinguishes between an initial period of "dreaming" during the struggle for independence, followed by the "euphoria and certainty" of the attainment of independence, and finally the "dysphoria and uncertainty" set off by the beginning of the civil war. Another possibility, advanced by Fonseca and Moreira, is to divide poetic production into "three great periods": the "rise to consciousness" of the 1950s to 70s, the "aesthetic innovations" of the 1970s, and the "eclecticism" and "discontinuity" of the 1980s. This perspective favours formal aspects over the mere engagement with the content and its relation to context (Fonseca / Moreira: Panorama, p. 20). The same periodization is adopted by Ana Mafalda Leite (Leite: Angola, pp. 142–164). Finally, Patrick Chabal proposes four phases that can be applied to all African literatures in the Portuguese language: an early "assimilation" period in which the European colonial culture still serves as a model; a "resistance" period in which the construction of an autonomous African culture becomes paramount; an "affirmation" phase that follows independence; and finally the current phase of consolidation of the national identity. See Patrick Chabal: Introduction. In: Id. / Angel / Brookshaw / Leite / Shaw (eds): *Postcolonial Literature*, pp. 1–28, here p. 10.

16 This conflation of the author and the poetic subject with a heterogeneous population is all but uncomplicated, an aspect that is enhanced in the Angolan case due to the profound divisions and conflicts within the population itself. Rather than claiming

First Movement: the Anti-colonial Struggle

The years of anti-colonial struggle, which started with the foundation of the resistance movement MPLA (People's Movement for the Liberation of Angola) in 1956 and continued until the attainment of independence in 1975, were a crucial phase in the establishment of the Angolan national identity. As indicated above, this was also a period of profuse cultural production, especially on the level of popular music and poetry. In their quest for fitting articulations of a specific Angolan experience, the poets resorted often to binary opposites that would set the Angolan people apart from the colonizing powers.

The sea revealed itself as an ideal trope to (re-)stage the drama of the brutal colonization and exploitation of the Angolan territory, placing the Angolans in the position of victims and the Portuguese in the role of perpetrators. Additionally, the emphasis on the transatlantic slave trade served as a means of exploring alternative connections beyond the Portuguese-Angolan axis. The Americas, to where the African slaves had been transported, acquired a new significance: instead of featuring simply as the point of arrival of the slaves—and hence a diaspora that remained separated from the original homeland by a vast ocean—they became a virtual meeting point, allowing the oppressed Angolans to enter into a dialogue with their oppressed distant relatives. At a time when struggles for liberation were becoming a global theme—from the Civil Rights movement of the USA; to the Communist revolution in Cuba; to the anti-colonial guerrillas in Africa—this newly discovered connection between the continents was indeed a valuable rhetorical means of exploring new ways of building the foundations of 'Angolanidade', a term commonly used to refer to the distinctive national character. While the ocean remained an obvious barrier between these different struggling peoples, poetry itself could be celebrated as a vehicle of communication, traversing geographic boundaries.

that we should read the poems as the ultimate expression of the prevalent opinions of the Angolan people, this analysis intends to highlight the instrumentalization of poetry for the propagation of a particular view or even ideology. In a way, this reflects the oversimplifications inherent in the construction of the imaginary community of the nation-state.

An early example of the notion of the sea as a 'contact zone' can be found in the poem "Soneto ao Mar Africano" by Geraldo Bessa Victor (1957):

> Ó grande Mar, que banhas estas plagas
> africanas, em ti ouço recados
> dum mundo a outro mundo, nos teus brados
> de prantos, risos, orações e pragas!
>
> Na dramática voz das tuas vagas,
> escuto os que, nos séculos passados,
> choraram nesse canto dos teus fados,
> cantaram nesse choro em que te alagas...
>
> Na tua voz eu ouço o Branco bravo,
> que semeou Portugal nestes recantos
> africanos, e ainda o Negro escravo
>
> – ao mesmo tempo indómito e servil –
> que regou com seu sangue e com seus prantos
> a semente fecunda do Brasil![17]

Opening with an apostrophe directed to the sea itself, the poet reveals an image of the Atlantic Ocean as a site of brutality in which black African people feature as the main sufferers. Mary Louise Pratt's famous definition of contact zones as "social spaces where cultures meet, clash, and grapple with each other, often in contexts of highly asymmetrical relations of power"[18] comes here fully to the fore. The imbalance of power permeates the poem and is especially visible in the rhymes 'bravo' / 'escravo' and 'servil' / 'Brasil' from the two last stanzas, in which the enslavement and servitude of the black African are made to contrast with the ferocity of the white Portuguese on the one hand and the financial success of Brazil on the other. These lines

17 "Oh great Sea, you who bathes these regions / of Africa, in you I hear messages / of one world to another world, in your screams / of cries, laughter, prayer and curses! / In the dramatic voice of your tides, / I hear those who, in centuries past, / cried in that singing of your *fados*, / sang in that crying where you flood yourself... / In your voice I hear the fierce White Man, / who planted Portugal in these corners / of Africa, and still the Black slave / – at once untamed and servile – / who irrigated with his blood and his tears / the fertile seed of Brazil!" Geraldo Bessa Victor: Soneto ao Mar Africano. In: Manuel Ferreira (ed.): *No Reino de Caliban: Antologia Panorâmica da Poesia Africana de Expressão Portuguesa*, vol. II: Angola, São Tomé e Príncipe. Lisboa: Seara Nova 1976, p. 58 (my translation).

18 Mary Louise Pratt: *Imperial Eyes: Travel Writing and Transculturation*. New York: Routledge 2008, p. 8.

emphasize the position of the African slave as a sacrificial victim and draw the reader's attention to the immense loss suffered by the African continent as a result of the slave trade—a loss that, significantly, constituted a gain for both Portugal and Brazil.

A noteworthy feature of the poem is the centrality of voices: the poetic subject serves as a mere vehicle for the uncanny sounds emanating from the sea. The slaves drowned in the Middle Passage or killed by the brutal conditions of the plantations come back to haunt the shores of Africa. They are, however, mixed with the voice of the white oppressors, whose influence has been so powerful that the very means of articulation of the African voices relies totally on the Portuguese culture. The poet enacts this first and foremost through the choice of poetic form: by opting for the sonnet, one of the defining markers of European literature, Bessa Victor plainly excludes a more 'authentic' African voice from the text. This textual exclusion is brought home by the rather ironic line "choraram nesse canto dos teus fados", in which even the crying of the enslaved Africans is colonized by the famous, traditionally Portuguese music style *fado*. The distinctly sad tone of *fado* music, which has frequently been used in Portuguese culture to address the sea and all its dangers, is here transposed to the African coast. While this allows the voices of the suppressed slaves to be heard, it also confines their mode of expression to the bounds of the colonized.

The theme of voices communicating across the ocean can also be found in "Mamã Negra (Canto da Esperança)" by Viriato da Cruz (1961). Due to its relative length, I quote only a few relevant passages from the poem:

> [...]
> Pela tua voz
> Vozes vindas dos canaviais dos arrozais dos cafezais
> dos seringais dos algodoais!...
> Vozes das plantações de Virgínia
> dos campos das Carolinas
> Alabama
> Cuba
> Brasil...
> [...]Vozes de Harlem Hill District South
> vozes das sanzalas!
> Vozes gemendo blues, subindo do Missisipi, ecoando

dos vagões!
Vozes chorando na voz de Corrothers:
Lord God, what will have we done
– Vozes de toda América! Vozes de toda África!
[...]
Pelos teus olhos, minha Mãe
Vejo oceanos de dor
Claridades de sol-posto, paisagens
[...]
mas vejo também que a luz roubada aos teus
olhos, ora esplende
demoniacamente tentadora – como a Certeza...
cintilantemente firme – como a Esperança...
em nós outros, teus filhos,
gerando, formando, anunciando –

o dia da humanidade

O DIA DA HUMANIDADE...[19]

Music obviously constitutes an important aspect of this poem, but instead of the laments of the *fado*, the poet opts to express the suffering of African slaves through the genre of African-American blues, which was so important for the affirmation of black identity in the USA. Also formally we notice greater openness, rhythmic variation, and a more confident appropriation of the Portuguese language that suggests a break from the colonizing culture.

Like Bessa Victor, Viriato da Cruz points to the broad outreach of the Angolan experience by linking the lyrical subject to other oppressed people in the Americas. The geographic scope has been widened here to include not only Brazil but also the USA and Cuba—which at the time were important sites of resistance, as mentioned above. This

19 "[...] Through your voice / Voices coming from the sugar, rice, coffee plantations / the rubber and cotton plantations!... / Voices from the plantations of Virginia / from the fields of the Carolinas / Alabama / Cuba / Brazil... / [...] Voices from Harlem Hill District South / voices from the villages! / Voices moaning the blues, / rising from the Mississipi, echoing / from the train wagons! / Voices crying in the voice of Corrothers: / *Lord God, what will have we done* / – Voices from all of America! Voices from all of Africa! / [...] Through your eyes, my Mother / I see oceans of pain / Clarities of a set sun, landscapes / Purple landscapes / [...] But I see (Oh! Indeed I see!...) / but I see also that the light stolen from / your eyes, sometimes shines / devilishly tempting – like Certainty... / glowingly firm – like Hope... / in us others, your children, / generating, forming, announcing – / the day of humanity / THE DAY OF HUMANITY!..." Viriato Da Cruz: Mamã Negra. In: Ferreira (ed.): *No Reino de Caliban*, pp. 173–174 (my translation).

shift of attention towards the revolutionary potential of the African diaspora is a crucial departure from Bessa Victor's sonnet. While the drama of the transatlantic slave trade is still very much at the core of the poem—epitomized in the phrase "oceans of pain", that serves as a title to this essay—"Mamã Negra" clearly displays a bolder tone that climaxes into the celebratory final lines. The sons of the black mother, a typical metaphor used to refer to the whole continent,[20] emerge here as the liberators of humanity. The silenced and sacrificed slaves ultimately triumph through the struggle of contemporary Africans and of the descendants of Africans in the Americas.

In spite of the tragedy at the origins of this unification, the sea features as a positive space, a contact zone that allows the non-conformist voices from different geographic areas to communicate with and eventually support each other. 1961 marked the beginning of the anti-colonial armed struggle in Angola, and poems such as "Mamã Negra" can be read as the battle cry of the resistance movement.[21] Within the few years that separate this poem from Bessa Victor's sonnet, there is a progression towards a more future-oriented utopian vision that corresponded to the Angolans' fight for independence. It was important here to avoid a sense of isolation and to inscribe the Angolan struggle within a broader context.

The combination of mourning for the sacrificed ancestors and celebration of a new beginning is also clearly visible in "Confiança" by Agostinho Neto (1979)[22], the nation's foremost poet, leader of the MPLA and the first president of independent Angola:

> O oceano separou-se de mim
> enquanto me fui esquecendo nos séculos
> e eis-me presente

20 In fact, Elleke Boehmer argues that postcolonial literature's depiction of Africa as a primal "motherland" has contributed to the perpetuation of gender roles in which men appear as active and visionary fighters, while women – metaphorically associated with the colonized motherland – assume the role of passive victims in need of protection. Elleke Boehmer: *Stories of Women: Gender and Narrative in the Postcolonial Nation.* Manchester: Manchester University Press 2005, pp. 22–41.

21 Like most poets discussed here, especially from this early period, Viriato da Cruz was member of the MPLA. He left the party in 1962, however, due to opinion clashes with its leader, Agostinho Neto.

22 The date indicates the publication of *Sagrada Esperança*, from which the poem is taken. This book is actually a collection of poems written by Neto in the 1950s and 60s.

reunindo em mim o espaço
condensando o tempo.

Na minha hisória
existe o paradoxo do homem disperso
Enquanto o sorriso brilhava
no canto de dor
e as mãos construíam mundos maravilhosos

John foi linchado
o irmão chicoteado nas costas nuas
a mulher amordaçada
e o filho continuou ignorante

E do drama intenso
duma vida imensa e útil
resultou a certeza

As minhas mãos colocaram pedras
nos alicerces do mundo
mereço o meu pedaço de chão.[23]

As in "Mamã Negra" we see here the indelible bond between the Angolan lyrical subject and the African diaspora in the USA, represented most clearly by the lynching of the character John. Neto begins by emphasizing the danger of memory loss for the Angolan people, presumably the state in which they found themselves under colonialism. It is only by realizing his painful link to the history of transatlantic slavery that the lyrical subject can actually attain full independence. The willful engagement with the ancestors' experiences of enforced dispersal, physical abuse and exploitation is vital for the construction of a new consciousness that actually empowers the subject—as the last stanza of the poem suggests. The indirect contribution of the Angolans for the wealth and well-being of nations in the Americas, a point that had already been touched upon by Bessa Victor, is presented here as an individual and direct achievement, and hence also a personal

23 "The ocean got separated from me / while I went on forgetting through the centuries / and here I am now / reuniting in me the space / compressing time / In my history / resides the paradox of the dispersed man / While the smile shone / upon the singing of pain / and the hands built wonderful worlds / John was lynched / his brother whipped on the naked back / his wife gagged / and the son remained ignorant / And from the intense drama / of an immense and useful life / arose the conviction / My hands have placed stones / on the world's foundations / I deserve my piece of ground." Agostinho Neto: *Sagrada Esperança*. Lisboa: Sá da Costa 1979, p. 79 (my translation).

sense of entitlement. The rhetorical *pars pro toto* turns every Angolan into a rightful owner of the national soil.

The war between Portugal and Angola, its most important African colony, lasted from 1961 to 1974. The high costs and the negative impact of this conflict on Portugal's international standing were a major contribution for the famous Carnation Revolution in 1974, the military coup that ended the dictatorship in Portugal. With it came the independence of the overseas colonies, including Angola. In a way, the utopian vision of "Mamã Negra" and the sense of entitlement of "Confiança" seemed to have become a reality.[24] Yet the repossession of the national territory proved to be a major challenge, as the country made a very shaky transition to independence: different liberation movements fought over governance, with support from the USA and the USSR, which sought to indirectly control the territory. By 1975, Angola had in fact plunged into a civil war. This would have a profound impact on the poetic production as well, and the sea would be used as a trope in quite a different manner.

Second Movement: the Civil War

Whereas the poetry of the anti-colonial struggle showed a manifest opening out across the ocean towards the Americas, during the protracted civil war we can witness an actual confinement of the poetic subject. Indeed, the discussion of poets of the 1980s and 90s in Angola often highlights the tendency towards introspection, subjectivism and experimentation with forms and themes.[25] The shattering effects of the armed conflict on the national imagination called for new ways of depicting the nation that were less oriented towards the simple binary opposites of the previous phase. Rather than being a vehicle for

24 It is worth stressing here that the MPLA, which rose to power upon independence, counted with extensive financial, logistic and ideological support from the USSR and from Cuba during and after the years of anti-colonial struggle. In this sense, rather than merely projecting an ideal of international cooperation, Angolan poets like Viriato da Cruz were essentially reflecting the reality of anti-colonial movements in Africa which became part of the greater matrix of the Cold War.

25 For a more detailed discussion of Angolan poetry of the 1980s, see Jorge Arrimar: A Nova Poesia Angolana In:*Zunâi* 4,21 (2010). http://www.revistazunai.com/materias_especiais/festival_tordesilhas2/comunicacao_jorge_arrimar.htm (accessed December 16, 2013).

international alliances, the sea now features more frequently as a barrier or even a threatening element, creating a sense of isolation.
The optimism of the previous phase clashes here with the violent reality of the civil war. The sea becomes a metaphor for the limited influence of humans over their fate, as the following excerpts from the poem "Mar Novo" by Manuel Rui (1976) convey:

> [...]2.
> Mas é novo este azul   tela rasgada
> é novo o nosso olhar.
> É nova esta forma gestual de espuma
> feita sabor de amor de guerra e de vitória
> em nossas bocas férteis em nossa pálpebras
> de antigo medo clandestino
> soletrando a lágrima
> quando era o nosso mar recordação também escravizada:
> caminho secular de ir e não vir.
>
> [...]6.
> Tudo é fugaz
> entre o desenho do teu pé na areia
> e a onda que desfaz
> a marca
> Entre a guerra e a paz
> retorno fisicamente o poema   a onda constante meditação primeira.
> Nós e as coisas.
> Nada permanece que não seja
> para a necessária mudança.
> Que o diga o mar.[26]

A sense of positivity prevails in part 2 of the poem, where the newness of the sea is contrasted with the loss of memory brought about by colonization, an image already met above in "Confiança". We can read in this celebration of newness an echo of the notion of the "new man" ("homem novo") which, under the influence of the MPLA's socialist ideology, became a frequent way of articulating Angola's turn

26 "[...] But it is new this blue torn canvas / new is our gaze. / New is this gestural form of foam / turned taste of love of war and of victory / in our fertile mouths in our eyelids / of ancient clandestine fear / spelling the tear / when our sea was memory also enslaved: / secular path of going and not coming. / [...] Everything is fleeting / between the outline of your foot in the sand / and the wave that erases / the traces / Between war and peace / return physically the poem the wave / constant first meditation. / Us and the things. / Nothing remains that is not for the necessary change. / Ask the sea." Manuel Rui: *Memória de Mar.* Lisboa: Edições 70 1980 (my translation).

towards a brighter future that would be constructed through labour and national unity.[27] Manuel Rui, himself a member of the MPLA, reflects this faith in the simple repetition of the term "new" in the first three lines of part 2. However, because the focus of this part is so strongly on the observer, there is a sense that the lyrical subject essentially projects onto the sea a series of impressions and associations. The poem thus reveals a gap between the perception of uniqueness and the actual reality of the sea, which remains inaccessible.

This ambiguity is reinforced in part 6, which closes the poem. The lyrical subject seems to negotiate between the horrors of the war and the need to believe in the "new man". Recognizing the ephemeral nature of existence itself and torn between the contingency of war and peace, the only certainty that the speaker can hold onto is the perpetual return of change. Like the wave, the observer's perception and the poem itself are in permanent transformation. The sea, which can erase traces of the past at will, becomes a symbol of transition—one that is beyond human control.

The exploration of the sea as an uncontainable force of nature can also be found in John Bella's "Noite [Ki]Anda" (1995), which relies heavily on Angolan myths, local habits and the Kimbundu register to convey the special relationship between the Angolan people and the Atlantic Ocean:

> Gritos do muzongue
> Confundem-se com viuvinhas na mulemba
>
> "Botessar" de feijão
> é dança circulante nos coqueiros
>
> Batucada de peixes na maré
> desbrava rede axilwanda
> Misanga da besangana
> "xinguilam" sob "calundús" de águas passadas
>
> Kyanda que anda da ponta ao cuidado
> procura Kalumba
> que deu wanga kalumba
>
> Na noite Kyanda

27 See Inácio Marques / Tatiana Pinto: Angola Independente: do *Homem Novo* à Crise Nitista. In: *Anais XI Luso Afro Brasileiro de Ciências Sociais. Diversidades e Desigualdades.* Universidade Federal da Bahia: Agosto 2011, http://www.xiconlab.eventos.dype.com.br/resources/anais/3/1308500913 (accessed July 1, 2013).

besa molho de ngana axilwanda
oferecendo kixilwanda à Kyanda[28]

At first sight the poem appears to be a re-enactment of rituals performed in honour of Kianda (as her name is more frequently spelled), goddess of the waters. Every November such rituals can be observed at the Ilha de Luanda, the famous beach strip off the shore of the capital frequented by locals and tourists. Essentially Kianda is venerated by fishermen with offerings of food, and these typical dishes ("muzongue", "botessar", "molho de ngana", "kixilwanda") are placed at the centre of Bella's poem: the sea is fed abundantly in an attempt to appease it. Its violent influence can be felt not only in the frenzy of the ritual itself, composed of drumming, dancing and screams, but also in the reappearance of the "calundús", the spirits of the ancestors, from the water. The traumatic memory of the slave trade once again emerges to haunt the living—but whereas in previous poems there had been a promise of liberation associated with this haunting, in "Noite [Ki]Anda" the hostile nature of the sea is highlighted by the death spell ("wanga") placed upon the city. In other words, rather than serving as a precious recollection of the sacrificed Angolans, the reemerging ancestors assume the shape of a threat.[29]

When this poem was written, death was not something of the past, buried in people's memories, but rather a permanent presence in the national imaginary. Just a few years earlier, in 1992, Luanda itself had been the site of a "bloodbath" when the results of the first democratic elections in the country plunged it back into armed conflict after a

28 "Screams from *muzongue* [fish stew] / Get mixed with whistling ducks on the mulemba [fig tree] / Bean "*Botessar*" [dish] / is a circling dance on the coconut trees / Drumming of fishes in the tide / breaks the net in Luanda / Glass beads of the young women / "Go into trance" under "spirits of ancestors" from waters past / *Kyanda* who walks from the tip to the caution / Searches for *Kalumba* [the Sea] / who put a death spell / In the Kyanda night [also 'the night that walks'] / Bless the gravy of mistress Luanda / offering *kixilwanda* [fish with manioc] to Kyanda." John Bella: *Água da vida*. Luanda: Delegação Provincial de Luanda da Cultura 1995, p. 32 (my translation).

29 Several Angolan texts of the late 20th and early 21st centuries have re-appropriated the myth of Kianda to express a similar unease about the developments in postcolonial Angola. Famous examples include Pepetela's novel *O Desejo de Kianda* (1995) and Manuel Rui's *Um anel na areia* (2002). See Carmen Secco: A Importância da Literature e das Artes Plásticas no Contexto da Cultura Angolana. In: *Fórum dos Angolanistas*, Universidade Estatal do Rio de Janeiro, September 2005. http://www.letras.ufrj.br/posverna/docentes/62671-1.pdf (accessed June 16, 2012).

short peace interregnum.[30] The violence of that episode and of the long civil war find a symbolic expression in Bella's depiction of the sea as a voracious deity intimately associated with death. This also gives the sea the appearance of an insurmountable barrier, like a cordon surrounding the city. The intractability of the sea is formally reinforced by Bella's decision to use Kimbundu terms to describe large parts of the action, making the poem inaccessible to anyone who does not possess knowledge of the language or the rituals. Thus also the sense of isolation of the Luanda inhabitants comes to the fore: at a time when the notion of national unity so important in the MPLA ideology was made meaningless by the prolonged civil war (often fought along ethnic and also linguistic lines), this focus on the local could be read as a questioning of the idea of the nation altogether.[31]

Both Manuel Rui and John Bella illustrate the departure of Angolan poets from the typical exaltation of liberation found in the first phase. Their emphasis is on the exploration of poetic language, including the metacritical level of reflection about the role of poetry itself. Another realm that is often explored in this phase is individual experience, which also contrasts sharply with the collective approach of the poems of the liberation struggle. The sea emerges here frequently as a means of expressing erotic love. However, placing the personal in the foreground does not preclude an engagement with the larger context, rather the opposite: the intimacy of individual experience becomes a powerful way of addressing the incomprehensible violence of the war. The poem "Resignação" by Maria Alexandra Dáskalos (1999) offers a fitting example:

> Da resignação nada sei.
> O mar está encapelado
> sou um barco.
> Guardo os sapatos, fecho as portas
> passeio à chuva.
> Espero o vento
> há que colher os frutos.

30 Birmingham: *Empire in Africa*, p. 161. Revisiting the city ten years later, Birmingham was shown the points of greatest violence during the 1992 confrontations: "At notorious intersections cars thought to contain opposition sympathizers had been stopped, the occupants shot, and the vehicles set ablaze." Ibid., p. 162.

31 Compare this especially with the poem by Viriato da Cruz, in which English is used in some verses to highlight the fraternal bonds between Angola and the USA. This move towards a global consciousness is totally undermined in Bella's poem.

Tu descansas serenamente
folha leve, por terra
fim de cacimbo

Os heróis não voltam.
Dormes, não queres estar vivo.[32]

The sea serves here as a mood-invested space, its fury reflecting the restlessness of the lyrical subject. The speaker's proclaimed refusal of resignation in the first line contrasts sharply with the absolute passivity of the lyrical thou in the final line. Underlying this tension between the two characters of the poem is the penultimate line "the heroes won't return". This could be read as a reference to the recent heroic history of the country, when Angolans fought for and attained independence, an achievement celebrated in the opening lines of the national anthem: "Oh fatherland, we will never forget the heroes of February 4th".[33] Yet in the poem the heroes seem to have been relegated to oblivion, incapable of inspiring the lyrical thou. The equation of sleep with death provides a striking expression of a lack of prospects. Unlike the poems of the first phase discussed here, there seems to be nothing in "Resignação" to look forward to. In spite of the opening line, the lyrical subject herself merely reacts to the natural elements and the seasons. Like a boat caught in a sea storm, she is waiting for change rather than pushing for it.

Third Movement: the Twenty-first Century

Between 1975 and 2002 half a million people were killed in the civil war, four million Angolans were displaced, and countless people (mainly women and children) were mutilated by landmines buried in the fields all over the country.[34] After 2002, Angola was able to enter

32 "Of resignation I know nothing. / The sea is restless / I am a boat. / I put away the shoes, close the doors / walk in the rain. / I wait for the wind / the fruits must be picked. / You rest serenely / light leaf, on the ground / end of *cacimbo* [cold season] / The heroes won't return. / You sleep, you do not want to be alive." Maria Alexandra Dáskalos: Resignação. In: Alice Brás / Armandina Maia (eds): *Vozes poéticas da lusofonia*. Sintra: Instituto Camões 1999, p. 35 (my translation).

33 February 4th 1961 marks the beginning of the armed struggle against the colonial occupation of Angola: on this day members of the MPLA attacked a number of government institutions in Luanda, including a military post, a prison and the radio station. Although the attackers failed to achieve their goal of liberating political prisoners, they caught international attention.

34 Angola is in fact known to have one of the world's highest concentrations of

a period of relative peace, but many tensions remain unresolved. In 2012 democratic elections were held for the second time in Angola's independent history, and the MPLA emerged again victorious. The country's president, José Eduardo dos Santos, leader of the MPLA, has now been in power for 33 consecutive years. Critical voices in the country and abroad have called attention to the high level of government corruption and repression,[35] an issue that has recently inspired civic activism and demonstrations.

The main topic of national and international contention has been the dramatic increase in Angola's wealth. Angola has become the second most important sub-Saharan supplier of crude oil, bringing more than 30 billion dollars of revenues in 2011 alone.[36] Luanda is now considered the second most expensive capital in the world, yet more than half of its 5 million inhabitants live in shantytowns that lack basic hygienic conditions. This sharp dissonance between extreme wealth and extreme poverty has had an impact also on the national imaginary.

In 2012, when the country was preparing for elections, the government-owned corporation Sonangol, which oversees petroleum and gas production, put out a full-page advertisement in national newspapers celebrating the 37th anniversary of Angola's independence. The ad shows a neatly lined up crowd of young black Angolans wearing the company's uniform, with an oil platform and a helicopter in the background. Only the faces of the workers in the first few lines are actually visible, while the majority forms literally an anonymous mass. Together, the helmets of this large mass make out the Angolan flag.

The ad establishes a direct link between petroleum and independence, while totally obliterating Angola's long history of armed conflict. The national flag stands as a symbol of patriotic pride and conveys a sense of perfect national unity, burying the memory of the civil war. The

anti-personnel mine victims. See Bruce Fish / Becky Durost-Fish: *Angola. 1880 to the Present: Slavery, Exploitation and Revolt.* New York: Chelsea House Publishers 2002, p. 134.

35 See for example Rafael Marques de Moraes: Growing Wealth, Shrinking Democracy. In: *New York Times*, August 29, 2012, p. 27; David Smith: Angola's human rights record. In: *The Guardian*, October 15, 2010, p. 4.

36 See Candido Mendes / Colin McLelland: Angola's Sonangol Earned $33.7 Billion in 2011 Oil Sales. In: *Bloomberg*, February 24, 2012. http://www.bloomberg.com/news/2012-02-24/sonangol-of-angola-earned-33-7-billion-from-oil-sales-last-year.html (accessed July 3, 2013).

heroes of yesteryear have been replaced by today's generation of young, entrepreneurial Angolans who collectively build a bright future by participating in the petroleum exploration. The MPLA's socialist rhetoric has been adapted to the country's new economic conditions, celebrating the Angolan people as a unified mass proudly contributing to the collective success and wealth.

Another striking detail of the ad is the prominent presence of the sea. The workers stand on a platform that triumphantly stretches over the high seas, creating a feeling of outward expansion. There is a realistic basis for this depiction: Angola's oil fields are in fact located offshore in the Atlantic Ocean. The very space that was once used to export African slaves, the continent's most valuable commodity, has now become the source of another valuable and globally coveted good, that somewhat ironically goes by the name of "black gold". In a perverse way, history appears to repeat itself. Accordingly, we find in the poetic discourse a treatment of the sea that bears significant resemblances to the years of anti-colonial struggle—adapting the tropes of that period to a new, postcolonial setting.

In the poem "Os meus pés descalços" by Décio Bettencourt Mateus (2006), Angola's newfound affluence is presented as a *fata morgana*. Following the lyrical subject's journey on foot, the country is depicted from different angles, revealing its lush landscapes and rich soil, sea, rivers and mines. The following excerpts, more directly related to the sea, illustrate the contrast between the speaker's destitution and the always visible, yet unattainable wealth:

> [...]
> Os meus pés empoeirados
> Acariciam subsolo rico, ouro negro a jorrar no alto mar
> Ouro negro a jorrar no offshore
> E no onshore
> Ouro negro a brotar
> Das entranhas do mar, para os meus pés esfomeados!
>
> [...] Os meus pés pescadores
> Banham-se em mares ricos
> Mares de garoupas, corvinas, carapau, mariscos...
> [...]
> Águas fartas a banharem os meus pés sofredores
>
> Os meus bolsos vazios
> Vêem outros bolsos vazios aterrar desnutridos

E depois, bolsos cheios
A levantar voo, a embarcar abastados
Bolsos cheios a embarcar com sorrisos
A embarcar abarrotados, oh que paraíso! [...][37]

The sense of entitlement that was so pronounced in Agostinho Neto's "Confiança" is here only implicit. The speaker is able to list the country's natural assets in great detail—with a whole stanza dedicated to the offshore oil wells; however, he remains largely detached, as if watching the described scenery through a screen. In spite of the direct contact between his feet and the water, the lyrical subject fails to explicitly express any notion that the wealth from the sea could belong also to him.

This is highlighted by the stanza describing the new migration movements in the country, which bring in more destitute people and let out the recently enriched. In the exclamation "oh que paraíso!" we can read perhaps admiration or even envy for the rich, but no actual resentment at the fundamental injustices witnessed by the speaker. The disempowerment of the lyrical subject is used by Bettencourt Mateus to actually highlight not only the unfair system of wealth distribution but also the general corruption behind the country's business transactions, symbolized by the synecdoche of the "bursting" pockets. Similar to the period of colonial exploitation, the sea is connected to an immense source of wealth that is essentially plundered by greedy individuals—only this time it is impossible to distinguish between outsiders (the colonizers) and insiders (the colonized). The drama of the nation is thus enhanced.

"Os meus pés descalços" testifies to the enduring power of Angolan poetry to articulate the country's problems and to stress inconsistencies in the dominant discourse. The poem can be read as a counterpoint to the triumphant narrative propagated by the MPLA government, as

37 "[...] My dusty feet / Caress rich soil, black gold spurting out of the high seas / Black gold spurting out on the offshore / And on the onshore / Black gold springing / From the bowels of the sea, towards my starving feet! / [...] My fisherman's feet / Bathe in rich seas / Seas full of bass, corvinas, mackerels, mollusks... / [...] Rich waters bathing my suffering feet [...] My empty pockets / See other empty pockets landing malnourished / And then, full pockets / Flying off, going on board packed / Full pockets going on board with smiles / Going on board bursting, oh what paradise! [...]." Décio Bettencourt Mateus: *Os meus pés descalços*, Luanda: União dos Escritores Angolanos 2008 (my translation).

seen in the advertisement above. It gives voice to the excluded who do not fit into the neat picture of a rising power. Since the beginning of the 21st century, however, a new form of lyrical expression has established itself alongside poetry, reaching a breadth of audience hitherto only dreamed of. As in most other countries, rap music now enjoys great popularity in Angola, and has become in fact one of the most common idioms of the younger generation. Here the impetus to speak out against injustices is even more pronounced, as hip hop remains the genre of popular music most associated with social critique.[38]

Perhaps the most (in)famous rapper in the country is MCK, whose latest album, *Proibido Ouvir Isto* (2012), sold 10,000 copies within 4 hours upon its release in Luanda.[39] The artist is distinguished by his uncompromisingly critical tone directed against the government (including often the president himself), corruption, and neo-colonial exploitation. Like many other rappers in the country,[40] he celebrates Angola's distinctive landscapes, traditional and urban cultures, and recent history, especially the period of the liberation struggle. Assuming the typical role of educator espoused by many rappers, MCK has made it his mission to readjust the nation's perception of its problems and opportunities.

In songs like "Políticas do Domínio" (2006), he criticizes the position of resignation adopted by many Angolans in the face of adversity, which is also expressed in the above poem by Décio Bettencourt Mateus. For MCK, the origins of this attitude lie deep in the colonial past, as the following excerpts from the song lyrics illustrate:

> O complexo de inferioridade controla o teu raciocínio
> Sinto nas tuas frases o vírus to extermínio
> Diagnóstico elaborado, políticas de domínio
> Vives loucamente apaixonado pelo Ocidente
> Desprezas a tradição oral do teu povo

38 See Tony Mitchell: Introduction. In: Id. (ed.): *Global Noise: Rap and Hip-Hop Outside the USA*. New York: Wesleyan 2002, pp. 1–38.

39 See Mário Lopes: Voz de Angola. http://www.buala.org/pt/palcos/voz-de-angola (accessed July 3, 2013).

40 The compilation *É Dreda Ser Angolano* (Lisboa: Rádio Fazuma 2007), which includes all the major artists, provides a good overview of the themes and styles in Angolan rap. It is important to highlight, however, that rap music remains very much an independent phenomenon, distributed and propagated through the Internet and underground channels.

> Ignoras a tua gente
> Estou consciente que não será fácil mudar o teu pensamento
> Este estúpido sentimento – não se jogam cinco séculos num dia
> O esquecimento – seja bem vindo a esta viagem ao passado
> Antes de eu nascer, cerca de 500 anos atrás no hemisfério do Sul
> Barcos do Atlântico desceram para África
> Homens de cabelos compridos e barbas longas
> Mais um mercado de aquisições humanas
> Estendia-se assim o império da civilização
> do norte ao sul, da costa ao interior
> [...]
> Conhece o verdadeiro eco da história de África
> Assume as viagens de descoberta da realidade trágica [...][41]

Similar to Neto's "Confiança", the lyrics denounce the Angolan people's loss of memory. But whereas in the first case the conditions for that loss were presented as essentially imposed by the colonial system, in "Políticas do Domínio" it is actually Angolans themselves who have failed to emancipate, as they choose to ignore their own origins to orient themselves towards the dominant Western culture. The lyrical subject proposes to unearth the buried past in order to reveal the roots of this submissiveness. Here a direct link is established between the present condition of the nation and the process of colonization that started 500 years earlier. Rather than repeating itself, we could say that history is presented here as an unbroken chain of events from which the Angolan subject emerges always as the exploited victim.
The sea is therefore imagined in analogous ways to the phase of anti-colonial struggle: it is the source of Angola's curse, the medium through which invaders reached the country and plundered its people. Ironically, the colonizers are presented as a caricature of God, sporting long hair and white beards—and indeed one of the points of criticism

41 "The inferiority complex controls your thinking / I feel in your words the virus of extermination / Elaborate diagnose, politics of domination / You live madly in love with the West / You despise the oral tradition of your people / You ignore your folk / I am conscious that it won't be easy to change your thinking / That stupid feeling – you don't discard five centuries in one day / Forgetfulness – welcome to this journey to the past / Before I was born, about 500 years ago in the southern hemisphere / Ships from the Atlantic came down to Africa / Men with white hair and long beards / Another market of human acquisitions / Thus the empire of civilization extended / from north to south, from the coast to the interior [...] / Meet the real echo of African history / Take on the discovery travels of the tragic reality [...]" MCK: *Nutrição Espiritual*. (Luanda: Masta K Produções 2006) (my transcription and translation).

in the lyrics is the way the Christian faith was instrumentalized to enslave the minds of the colonized people. As a reenactment of that first tragic moment of contact between oppressors from the northern hemisphere and victims from the southern hemisphere, the song offers a postcolonial rewriting of Europe's glorious history of transatlantic travels, applying the famous trope of the "discoveries" to a new context that should in fact lead to the emancipation of the Angolan consciousness.

A similar treatment of the long-term negative effects of colonialism can be found in the work of Bob da Rage Sense, one of the most well-known Angolan expatriate rappers. Like MCK, Bob is vociferously critical of the current political and financial situation in Angola. His suggestively entitled album *M.P.L.A.* (2007) subverts the famous acronym of the governing party by translating the initials into: "Menos Pão, Luz, Água" (Less Bread, Light, Water). These have been common grievances throughout Angola at least since independence. The self-appointed vehicle for emancipation and progress, the MPLA, is here turned into the very source of ruin of the Angolan people. The rapper, who presents himself as a leftist poet, has in fact appropriated many of the symbols of struggle and revolution from the 1960s, in order to apply them to the current situation of Angola.

In the song "P.A.L.O.P." (2007), Bob again produces a pun based on an acronym: the original official meaning of "Países Africanos de Língua Oficial Portuguesa" (African Countries with Portuguese as Official Language), used to identify Portugal's ex-colonies in Africa, is turned into the offensive "Pretos Africanos de Língua Obrigatória Portuguesa" (Black Africans with Portuguese as Imposed Language). The lyrics deal with a topic that is familiar to hundreds of thousands of Angolans today, namely emigration. The capital of Portugal, where Bob da Rage Sense lives, has become an important centre for Angolan expatriates since the outbreak of the civil war. As a contact zone, the city reveals the same power imbalance that was experienced by Africans during the transatlantic slave trade. Precisely this link between past and present is stressed in the following excerpt of the song:

> [...]Palop, o vestígio kolonial ke ainda hoje se vê
> Angola, Moçambique, Cabo-verde, São-tomé e Guiné
> foi-nos dada uma religião racista, foi-nos dada a prisão
> foi-nos dada a eskravidão e a justifikação foi a expansão

comercial, atravessámos oceanos
muitos p'ró mar nos atirámos, afogámos
impuseram-nos uma kultura ke não era nossa
disseram ser a civilização mas era só vossa
proibiram o kimbundo o tchokwe e o umbundo
obrigaram a falar o português e enkarar um novo mundo
foram 500 anos, inkontáveis são os danos
e loukuras kausadas pelas turturas aos afrikanos
dividiram reinos pretos já existentes
dizimaram sem piedade todos os resistentes
aonde estão os descendentes e suas gentes
as chagas nas kostas dos velhos hoje são evidentes

kuando a independencia xegou nada mudou
o afrikano luso-obrigado emigrou e o ke enkontrou
um gueto kriminoso no kual mergulhou
nesta sociedade desigual a eskravatura nunka akabou
enkuanto afro-descendentes alimentam uma falsa esperança
o português ofereceu-nos o inferno komo herança
muitos entregam-se ao krime komo vingança
mas o kriminoso real ke nos oprime numa mansão deskansa

[Refrão:]
Eu sou, tú és, ele é....eskravo do passado
Nós somos, vois sois, eles são... a razão deste fado[42]

In a move analogous to Viriato da Cruz' "Mamã Negra", the song lyrics transcend the Angolan experience to establish a connecting link with other people—this time not across the Atlantic, but within Africa

42 "[...] Palop, the colonial traces still visible today / Angola, Mozambique, Cape-Verde, San-Tomé and Guinea [Bissau] / we were given a racist religion, we were given prison / we were given slavery and the justification was the expansion / of commerce, we crossed oceans, many of us jumped into the sea, drowned / they imposed a culture that wasn't ours / they called it civilization but it was only yours / they forbade kimbundo, tchokwe, umbundo / they made us speak Portuguese and face a new world / 500 years it lasted, countless damage / and madness caused by the torture imposed on Africans / they divided already existing black kingdoms / ruthlessly annihilated all resistance / where are the descendants and their people / the scars on the backs of the elders are still evident / when independence came nothing changed / the Luso-imposed African emigrated and what did he find / a criminal ghetto into which he dove / in this unequal society slavery has never ceased / while afro-descendants feed on a false hope / the Portuguese offered us hell as legacy / many have turned to crime in revenge / but the real criminal who oppresses us lives at ease in a mansion / [Chorus:] I am, you are, he is... / Slave of the past / We are, you are, they are... / The reason for this *fado* / fate." Bob da Rage Sense: *MPLA (Menos Pão, Luz, Água)* (Lisboa: Footmovin' (Bomberjack) 2007=. Lyrics available at http://letras.com/bob-da-rage-sense/973378/ (accessed May 19, 2012) (my translation).

itself. The close bond between Africans colonized by the Portuguese also goes back to the transatlantic slave trade, which is here depicted through emblematic images of the Middle Passage. The point of arrival is deemed less important than the tragic process of departure from the homeland, emphasizing the collective trauma. The rapid transition between verses produces startling associations: the word "drowned", which refers to those who voluntarily jumped to the sea and chose death over a life of slavery on the other side of the Atlantic, can also be applied to the elimination of national cultures in the following line. These were figuratively drowned by the colonial imposition of a new language and new beliefs.[43] This link between the sea and the destruction of the African identity is further applied to the present, as the descendants of the culturally enslaved Africans are forced to emigrate in search for better living conditions—in a way replicating the departure of their ancestors—only to end up "diving" into the violence and criminality of the urban ghettos in the ex-metropolis.

The devastating experience that still haunts today's Africans is encapsulated in the phrase "slaves of the past" in the song's chorus. As in MCK's "Políticas do Domínio", Bob suggests that there has essentially been no change in the 500 years since the Portuguese landed on the African shores. The lyrics not only thematize this enslavement, but also perform it linguistically. The chorus mockingly replicates the conjugation of the verb "to be", which is a typical exercise in the acquisition of a new language. For the lyrical subject, the Portuguese language remains a bitter reminder of the persistence of colonial subjugation—hence also the effort of subverting it through an 'Africanized' spelling (the abundance of the letter 'k' is reminiscent of the Kimbundu language) as well as through the irony of the refrain. That the language carries also a whole cultural baggage can be seen in the final pun of the chorus, using the word "fado", which in Portuguese can either mean "fate" or refer to the typical music style *fado*. Since the message is delivered in the form of a song, the intended meaning of this word remains highly ambiguous.

43 It should be noted that, although the speaker addresses all of Portugal's ex-colonies in Africa, when it comes to the suppressed languages he actually refers only to those spoken within the Angolan territory. His focal point remains Angola's national culture and identity.

Very much like Geraldo Bessa Victor's "Soneto ao Mar Angolano", the lyrics of "P.A.L.O.P." (re)enact a forceful experience of colonization from which the African speaker can no longer free himself, as it has become part of his cultural identity. Our analysis thus turns full circle. In spite of the 50 years separating the two texts, the striking parallels suggest that the drama of the Angolan experience is far from overcome. Even though the outward circumstances have changed significantly, the essential exploitation of the nation has remained unchanged.

Conclusion: Voices and Echoes from the Atlantic

The value of metaphors, Laura Murphy suggests, is that they "package a traumatic past in a way that is digestible, communicable and comprehensible. They turn literal pain into images that transmit the memory of suffering through generations."[44] This applies wholly to Angola's national imaginary. As this analysis has shown, the Atlantic Ocean has been central for the construction, transformation and adaptation of the country's national identity, because it is so intimately associated with the experience of the slave trade. While this event has endured in the nation's cultural memory, its translation into poetic discourse has been wide-ranging, as authors adapted it to new contexts. Another important aspect of metaphors is in fact their flexibility. As Murphy argues, "images of slave trade violence" can be "[layered] onto scenes of contemporary political corruption, social tensions and community cruelty".[45] In the specific case of Angola, in the second half of the 20th and early 21st centuries, metaphors of the sea have been used to articulate the discontent of three consecutive generations: those who fought for independence from the colonial power, those who faced the disenchantment and horrors of the civil war, and those who witness the ongoing corruption and exploitation of post-war Angola.

In the first phase discussed here, the sea is often associated with the dispersal of Africans and hence the "dismemberment" of an imagined unitary African identity, to use Ngugi Wa Thiong'o's formulation. In their effort to celebrate an emerging national pride, the poets at the same time emphasized Angola's vital contribution to the wealth of

44 Laura T. Murphy: *Metaphor and the Slave Trade in West African Literature*. Athens, OH: Ohio Press University 2012, p. 21.

45 Ibid.

the Americas. While this was intended to express the tragedy of the enslavement of Africans, it also served to announce the possibility of a new beginning, as the Angolan people gained consciousness of their worth (quite literally) and thereby gathered enough strength to oust their oppressors, with the support of their distant relatives across the Atlantic. The metaphor of the voices emanating from the sea was a powerful expression of the Angolan battle cry against colonialism.

The sense of triumph over oppression was short-lived, though, and the second phase of poetic production, during the civil war, applied images of the sea in a distinctive way that reflected the experiences of the poets. Metaphors of destruction and loss of control are prominent in the poems discussed here. The sea is no longer a space of communication, but a—sometimes dangerous—barrier. The sense of isolation and vulnerability functions as an indirect expression of the doubts about the postcolonial project of national unity in the face of a devastating civil conflict.

While the end of the civil war in the early twenty-first century inspired the official discourse to celebrate the nation again as a source of collective pride, the poetic discourse has often been less optimistic. Here the sea features again as a space where plundering and treachery abound. As in the first phase, the speakers essentially contemplate the loss of identity and wealth that seem to be sucked away by the sea itself. Most importantly, the Atlantic Ocean becomes a metaphor for an oppressive history that is anything but overcome—quite the opposite, the 500 years of colonial exploitation are like an unmovable block, barring the way towards true liberation. The project of the anti-colonial struggle is therefore far from completed.

These echoes of the revolutionary rhetoric and poetic tropes of the 1950s and 60s in the twenty-first century deserve some attention. Especially Angolan rappers have been prone to emulate and celebrate the country's founding ideals, which were rooted in the international anti-colonial discourse of the 1960s. The same values of independence, redistribution of wealth and celebration of local cultures and identities find an expression in the rappers' fiercely critical attitude towards the nation. Born after independence and raised during the civil war, many of the poets and rappers currently active in Angola were exposed to the socialist rhetoric of the MPLA, which promised "everything for the people", as one of its slogans announced. The war itself was seen as a deterrent to actual economic bliss. Once peace was established, however, the promises remained unfulfilled.

The newly established chasm between wealth and poverty provides insurmountable evidence of ongoing injustices and serves as fuel for the younger generation's dissatisfaction. In their commitment to the imagined community of Angola, they take on the role of corrective earlier assumed by the poets of the 1950s and 60s, and point to the shocking continuity between colonial and neocolonial exploitation of the nation. The sea, with its vast qualities as a memory site, has revealed itself as the most constant metaphor to express this ongoing drama.

Bibliography

Anderson, Benedict: *Imagined Communities: Reflections on the Origin and Spread of Nationalism.* London: Verso 1991.

Angolan Rap: When Money Stops Talking. In: *The Economist*, March 31, 2012, http://www.economist.com/node/21551547 (accessed July 1, 2013).

Arrimar, Jorge: A Nova Poesia Angolana. In: *Zunái* 4,21 (2010). http://www.revistazunai.com/materias_especiais/festival_tordesilhas2/comunicacao_jorge_arrimar.htm (accessed December 16, 2013).

Bella, John: *Água da vida.* Luanda: Delegação Provincial de Luanda da Cultura 1995.

Bender, Gerald J.: *Angola Under the Portuguese: The Myth and the Reality.* Trenton, NJ: Africa World Press 2004.

Birmingham, David: *Portugal and Africa.* Athens, OH: Ohio University Press 1999.

—: *Empire in Africa: Angola and Its Neighbours.* Athens, OH: Ohio University Press 2006.

Boehmer, Elleke: *Stories of Women: Gender and Narrative in the Postcolonial Nation.* Manchester: Manchester University Press 2005.

Chabal, Patrick: Introduction. In: Id. / Moema Parente Angel / David Brookshaw / Ana Mafalda Leite / Caroline Shaw (eds): *The Postcolonial Literature of Lusophone Africa.* London: Hurst & Company 1996.

Corrado, Jacopo: The Rise of a New Consciousness: Early Euro-African Voices of Dissent in Colonial Angola. In: *e-JPH* 5,2 (Winter 2007). http://www.brown.edu/Departments/Portuguese_Brazilian_Studies/ejph/html/issue10/html/jcorrado_main.html (accessed July 1, 2013).

—: *The Creole Elite and the Rise of Angolan Protonationalism: 1870–1920.* New York: Cambria Press 2008.

Da Cruz, Viriato: Mamã Negra. In: Ferreira, Manuel (ed.): *No Reino de Caliban: Antologia Panorâmica da Poesia Africana de Expressão Portuguesa*, vol. II: Angola, São Tomé e Príncipe. Lisboa: Seara Nova 1976, pp. 173–174.

Dáskalos, Maria Alexandra: Resignação. In: Alice Brás / Armandina Maia (eds): *Vozes poéticas da lusofonia.* Sintra: Instituto Camões 1999, p. 35.

Fish, Bruce / Becky Durost-Fish: *Angola. 1880 to the Present: Slavery, Exploitation and Revolt.* New York: Chelsea House Publishers 2002.

Fonseca, Maria Nazareth Soares / Moreira, Terezinha Taborda: Panorama das Literaturas Africanas de Língua Portuguesa. *Cadernos CESPUC de Pesquisa, Série Ensaios, n. 16: Literaturas Africanas de Língua Portuguesa.* Belo Horizonte: Set. 2007. http://www.pucminas.br/raca_cor_etnia/ppt/texto_terezinha.pdf (accessed July 1, 2013).

Hochschild, Adam: *King Leopold's Ghost: A Story of Greed, Terror and Heroism in Colonial Africa.* New York: Houghton Mifflin Company 1998.

Leite, Ana Mafalda: Angola. In: Ead. / Patrick Chabal / Moema Parente Angel / David Brookshaw / Caroline Shaw (eds): *The Postcolonial Literature of Lusophone Africa*. London: Hurst & Company 1996, pp. 103–164.

Lindsay, Lisa A.: *Captives as Commodities: The Transatlantic Slave Trade*. Upper Saddle River, NJ: Pearson / Prentice Hall 2008.

Marques, Inácio / Tatiana Pinto: Angola Independente: do *Homem Novo* à Crise Nitista. In: *Anais XI Luso Afro Brasileiro de Ciências Sociais. Diversidades e Desigualdades*. Universidade Federal da Bahia: Agosto 2011. http://www.xiconlab.eventos.dype.com.br/resources/anais/3/1308500913 (accessed July 1, 2013).

Mateus, Décio Bettencourt: *Os meus pés descalços*, Luanda: União dos Escritores Angolanos 2008.

Mendes, Candido / Colin McLelland: Angola's Sonangol Earned $33.7 Billion in 2011 Oil Sales. In: *Bloomberg*, February 24, 2012. http://www.bloomberg.com/news/2012-02-24/sonangol-of-angola-earned-33-7-billion-from-oil-sales-last-year.html (accessed July 3, 2013).

Mitchell, Tony: Introduction. In: Id. (ed.): *Global Noise: Rap and Hip-Hop Outside the USA*. New York: Wesleyan 2002, pp. 1–38.

De Moraes, Rafael Marques: Growing Wealth, Shrinking Democracy. In: *New York Times*, August 29, 2012, p. 27.

Murphy, Laura T.: *Metaphor and the Slave Trade in West African Literature*. Athens, OH: Ohio Press University 2012.

Neto, Agostinho: *Sagrada Esperança*. Lisboa: Sá da Costa 1979.

Ngugi Wa Thiong'o: *Something Torn and New: An African Renaissance*. New York: Basic Books 2009.

Padilha, Laura Cavalcante: Guerra, poesia, estilhaç[ament]os - um olhar para Angola. *Mulemba* 1 (2009), pp. 1–14.

Pawson, Laura: Angola is stirred by the spirit of revolution. In: *The Guardian*, March 8, 2011, http://www.guardian.co.uk/commentisfree/2011/mar/08/angola-spirit-revolution (accessed July 1, 2013).

Polgreen, Lydia: Change Unlikely From Angolan Election, but Discontent Simmers. In: *New York Times*, August 31, 2012, p. 11.

Pratt, Mary Louise: *Imperial Eyes: Travel Writing and Transculturation*. New York: Routledge 2008.

Rui, Manuel: *Memória de Mar*, Lisboa: Edições 70 1980.

Secco, Carmen: A Importância da Literature e das Artes Plásticas no Contexto da Cultura Angolana. In: *Fórum dos Angolanistas*, Universidade Estatal do Rio de Janeiro, September 2005. http://www.letras.ufrj.br/posverna/docentes/62671-1.pdf (accessed June 16, 2012).

Smith, David: Angola's Human Rights Tecord. In: *The Guardian*, October 15, 2010, p. 4.

Victor, Geraldo Bessa: Soneto ao Mar Africano. In: Manuel Ferreira (ed.): *No Reino de Caliban: Antologia Panorâmica da Poesia Africana de Expressão Portuguesa*, vol. II: Angola, São Tomé e Príncipe. Lisboa: Seara Nova 1976, p. 58.

Discography

Bob da Rage Sense: *MPLA (Menos Pão, Luz, Água* (Lisboa: Footmovin' (Bomberjack) 2007).

É Dreda Ser Angolano (Lisboa: Rádio Fazuma 2007).

MCK: *Nutrição Espiritual.* (Luanda: Masta K Produções 2006).

Oceanic Modernity in Contemporary Narratives – Remembering Slavery in Brazil and Angola

Ineke Phaf-Rheinberger

1. Introduction

Homi Bhabha claims in *The Location of Culture*[1] that, with the transition to the twentieth-first century, modernity has to be reviewed. For this purpose, he proposes to look at a "new" tendency in art and literature, focusing on migration in past and present. The publication of Bhabha's book was a mayor event; it meant the shift away from his focus on *Nation and Narration*, his earlier edited book, in which he offered "theoretical perspectives on the representation of cultural difference, both in the establishment of 'old' national communities, and in the contemporary context of transnational migration and cultural displacement".[2] *The Location of Culture* was a significant step toward the discussion of globalization in cultural studies and Bhabha's claim has been taken up since then, pointing to transatlantic migrations, diasporas, transnationality, hybridity, and the circulation of knowledge in general. However, I would like to emphasize a specific aspect of Bhabha's argumentation. Little attention has been paid to the fact that he emphasizes foundational borderline functions for the dynamics of these global moves:

* I am grateful to Michael Mann and Seth Schindler for their careful readings and critical comments on my essay.

1 Homi K. Bhabha: *The Location of Culture*. London: Routledge 1994.

2 Flap text on dust jacket of Homi K. Bhabha (ed.): *Nation and Narration*. London: Routledge 1990.

> "The harbour is the site in which material goods appear in bulk, in the very flux of exchange." The harbour and the stockmarket become the *paysage moralisé* of a containerized, computerized world of global trade. Yet, the non-synchronous time-space of translational "exchange", and exploitation, is embodied in a navigational allegory.[3]

Bhabha "locates" the non-synchronous time-space of "translational exchange" in port cities, conceived as foremost global agents due to their air of mobility as the title of the special issue of *Comparativ* suggests.[4] Mobility is characteristic for urban modernity in capitalist societies and conveys the impression of a continuous transformation or transition from one to another situation, created by the tension formulated in Arthur Rimbaud's verse: "Il faut être absolument moderne",[5] Henri Lefebvre's *Introduction to Modernity*[6] and *Writing on Cities*,[7] or Marshall Berman's description of the experience of modernity in *All That is Solid Melts into Air*.[8]

Taking these reflections as my point of departure, the first argument in this essay is that this global move is addressed in contemporary narratives from Brazil and Angola, in relationship to the remembrance of slavery. Without doubt, the countries' port cities were trans-shipment centers and also quite liberal places for cosmopolitan ideas and thoughts.[9] Consequently, for Brazil and Angola, Salvador da Bahia and Luanda,[10] linked to one another through the oceanic interspaces, are crucial places. Many titles of works by Brazilian historians and literary criticism on Africa emphasize the ocean's role in the constitution of

3 Bhabha: *The Location of Culture*, p. 8.

4 Lars Amenda / Malte Fuhrmann (eds): Hafenstädte: Mobilität, Migration, Globalisierung (= Special Issue, *Comparativ. Zeitschrift für Globalgeschichte und vergleichende Gesellschaftsforschung*, 2, 17 (2007)).

5 Arthur Rimbaud: Adieu. In: Id: *Sämtliche Dichtungen. Französisch und Deutsch*, ed. and trans. by Walther Küchler. Heidelberg: Schneider 1982, pp. 322–325, here p. 32.

6 Henri Lefebvre: *Introduction to Modernity: Twelve Preludes*. September 1959–1961, trans. by J. Moore. London: Verso 1995 [French, 1962].

7 Henri Lefebvre: *Writing on Cities*, ed. and trans. by E. Kofman / E. Lebas. Oxford: Basil Blackwell 1996.

8 Marshall Berman: *All that is Solid Melts into Air. The Experience of Modernity*. London: Verso 1983.

9 Carolyn Cartier: Cosmopolitism and the Maritime World City. In: *The Geographical Review* 89,2 (April 1999), pp. 278–289, here p. 280.

10 Salvador was the urban center of the Brazil until 1763, when Rio de Janeiro took over until Brazilian independence in 1889. Angola became an independent republic in 1975.

the South-South axis: *The formation of Brazil in the South Atlantic in the Sixteenth and Seventeenth Century*[11]; *A River Called the Atlantic*[12]; and *As if the Sea Would Be a Lie.*[13] In her interview with Luiz Felipe de Alencastro about his historical study *O trato dos viventes. A formaçãcao do Brasil no Atlântico Sul* (the treatment of the living), Mariluce Moura states:

> The central point in this book is undoubtedly the view that the roots of this nation lie, not within its own border, but in a Portuguese-Brazilian and Portuguese-African transcontinental space, strongly sustained by an economic zone consisting of Brazil and Angola, which remained in place from the sixteenth century up until the end of the slave trade in 1850.[14]

The pivotal role of Salvador and Luanda brings up the second and equally important argument about to what extent this "oceanic" dimension contributes to the construction of modernity in the urban environments. The city maps show that both cities are modeled as having an up- and a downtown city next to a large bay giving a haven to all types of ships. Bhabha conceives such spaces as the gathering place for the passage "of the massive economic and political diaspora of the modern world", as an aspect that has been "blasted out of the continuum of history".[15] The labor historians Donna Gabaccia and Dirk Hoerder have indirectly responded to Bhabha's claim by paying attention to this diaspora phenomenon.[16] They formulate the need for comparative research on flows of people who migrate overseas in their—voluntary or forced—search for employment and propose including them in the continuum of historical research. According to my argument, these sea-borne migrations contribute decisively to modernity and mobility in reference to gender relations, naming, life style, social status, intertextualities, and historical memory. In this

11 Luiz Felipe de Alencastro: *O trato dos viventes. A formação do Brasil no Atlântico Sul. Séculos XVI e XVII.* São Paulo: Companhia das Letras 2000.

12 Alberto da Costa Silva: *Um rio chamado Atlântico. A África no Brasil e o Brasil na África.* Rio de Janeiro: Edições Nova Fronteira / UFRJ 2003.

13 Rita Chaves / Carmen Secco / Tania Macêdo (eds): *Brasil / África. Como se o mar fosse mentira.* São Paulo: Editora UNESP 2006.

14 Mariluce Moura: Luiz Felipe de Alencastro: Brazil's observer in the South Atlantic. http://revistapesquisa.fapesp.br/en/2011/10/01/luiz-felipe-de-alencastro-brazils-observer-in-the-south-atlantic (accessed October 14, 2013).

15 Bhabha: *The Location of Culture*, p. 8.

16 Donna R. Gabaccia / Dirk Hoerder (eds): *Connecting Seas and Connected Ocean Rims. Indian, Atlantic, and Pacific Oceans and China Seas Migrations from the 1830s to the 1930s.* Leiden: Brill 2011.

way, the slave trade from Africa to the Americas—Paul Gilroy's Black Atlantic—will be integrated into a global picture and I will explore how this "oceanic modernity"[17] is constructed in contemporary narratives from Brazil and Angola.

2. Salvador da Bahia

At a colloquium of the Jorge Amado Foundation dedicated to the novel *Mar Morto* (dead sea)[18] in 2008, Helio Pólvora posed the question of the absence of the sea in Brazilian literature, though it deals with a country whose oceanic coast constitutes a significant part of its frontier.[19] Because of being the exception to this absence, Pólvora argues that *Mar Morto* is an outstanding work, being the most lyrical work of the legendary Jorge Amado, also called the Brazilian Victor Hugo; in Amado's novel the sea is a character in itself. Amado certainly knew Victor Hugo's *Les travailleurs de la mer*, translated into Portuguese by Joaquim Maria Machado de Assis as *Os Trabalhadores do mar*,[20] which Hugo dramatizes the struggle of an experienced fisherman against the forces of the sea, symbolizing danger and death. His main character, at the end, commits suicide in the English Channel. Amado's main character, in his turn, is an experienced fisherman from Salvador da Bahia, who lives with the sea in a symbiosis: "The sea becomes Yemajá's home, a goddess, transforming herself into a Brazilian and liquid Olympus."[21] For Amado's fisherman, Yemayá is as important as Lívia, the woman he shares his life with in the city. The novel is concerned with the tension between the modernization of urban life in Salvador, on the one hand, and, on the other, people like the fisherman who traditionally

17 After having coined this term for my research, I discovered in the Internet that the term also was used as the title of a conference on architecture in New Taipei City, Taiwan, on 5–6 January 2013: *Ocean Modernity. The Asia-Pacific Tectonic Culture and Postcolonialism.* The conference was organized by the Taiwanese anthropologist and cultural politician Chi-Nan Chen and addressed architecture and urban communities.

18 Jorge Amado: *Mar morto*, posfácio de Ana Maria Machado. São Paulo: Companhia das Letras 2008 [1936].

19 Hélio Pólvora: Jorge Amado e o romance do mar. In: *Colóquio Jorge Amado. 70 anos de Mar morto.* Salvador da Bahia: Fundação Jorge Amado 2008, pp. 53–60.

20 Victor Hugo: *Os trabalhadores do mar*, trans. Joaquim Maria Machado de Assis. São Paulo: Editora Nova Cultural 2002 [1866].

21 "O mar passa a ser a morada de Iemanjá, uma deusa, transformando-se num olimpo brasileiro e liquid." Ana Maria Machado: Pósfacio. A invenção da Bahia. In: Amado: *Mar Morto*, pp 275–280, here p. 278. (All translations from Portuguese into English in this essay are by its author, except where indicated otherwise).

live from and with the sea. This way of life is likely to disappear, as is illustrated by the main character's death in the sea at the end.

In Amado's novel, Yemayá is the personification of the link with Africa, having made her way from the African coast to Brazil.[22] She is "Iemanjá, Mistress of the seas and the saveiros",[23] the *saveiro* being the typical wooden boat of the fishermen in the Reconcavo, the bay area of Salvador da Bahia, of whom Amado's fisherman is a representative. Amado was friends with the musician Dorival Caymmi, who was inspired by *Mar morto* to compose his song "É doce morrer no mar" (it is sweet to die in the sea), famous in Brazil as well as in lusophone Africa.

During the same symposium, Caymmi's daughter Stella gave the details of the composition of this song when her father lunched with Amado and other friends.[24] Mia Couto, the well-known writer from Mozambique, who spoke in Salvador three years later on the occasion of a Amado symposium with a speech called "Sea and Life", pointed out that the Caymmi / Amado song impressed him so deeply as a boy that he thought that ocean waves were green, because the song's lyrics said so. When he was young, growing up in the Portuguese overseas province Mozambique, reading Amado's books were an awakening, because this Brazilian author introduced black people, until then a rare appearance in the literature of the metropolis. Couto therefore judges Amado's sensuality and religiosity as being close to African realities and argues that they inspired the feeling that changes were possible at the very moment that the Portuguese dictatorship imposed its rigid political system. Couto also claims that the sea created by Jorge Amado is not only a geographical one, but also a sea "within us, the sea that is alive within our lives".[25]

Obviously, Amado's sea—the sea that lives within us—is indissolubly linked to Salvador da Bahia and Ana Maria Machado writes in her

22 Félix Ayoh' OMIDIRE: La identidad frente al poder. La asimetría de Yemayá en África y América Latina. In: Ineke Phaf-Rheinberger (ed.): *Historias enredadas. Representaciones asimétricas con vista al Atlántico*. Berlin: Ed. Tranvía 2011, pp. 145–165.

23 "Iemanjá, Dona dos mares e dos saveiros." Amado: *Mar Morto*, p. 11.

24 Stella Caymmi: A parceria musical de Dorival Caymmi e Jorge Amado na adaptação teatral de *Terras do sem fim*. In: *Colóquio Jorge Amado*, pp. 77–94.

25 Mia Couto: O mar que Jorge Amado criou não é o que banha no sentido geográfico, é mar que está dentro de nós, o mar que está vivo dentro de nossas vidas. http://www.expressodasilhas.sapo.cv/pt/noticias/go/-mia-couto (accessed October 14, 2013).

postscript to *Mar morto* that it is an "urban" sea.[26] The novel addresses a city that was Brazil's administrative and cultural centre, as well as the doorway to other places on the globe. Yemayá also has other names, such as Janaína and Aícoa,[27] and is said to have five names all together. Amado was not the first author to introduce Yemayá in contemporary Brazilian literature. He shares this premiere with his stylistic antipode, the *modernismo* writer Mário de Andrade, the author of *Macunaíma*, a novel in the form of a rhapsody describing the lack of any identifiable character of the Brazilian people. Andrade mentions Yemayá in his seventh chapter on a Macumba ceremony in Rio de Janeiro, in which citizens from all social levels of local society participate.[28] In *Macunaíma,* however, Yemayá remains merely a marginal presence, in order to underline that Macunaíma hosts different religions in Brazil. In *Mar morto*, in contrast, the presence of Yemayá is felt everywhere—astonishing in a novel by an author labelled a social realist in that first period of his productive life.

3. The City of Yemajá

Jorge Amado's Salvador da Bahia has greatly influenced a contemporary Brazilian author, Ana Maria Gonçalves, who has elaborated a life cycle around Yemayá in her voluminous novel *Um defeito de cor* (a deficiency of color).[29] Although it counts 949 pages, this book was already in its seventh edition in 2011. It is Gonçalves' second novel[30] and she remarks in her prologue how she conceived the idea of writing this book. When looking for literature about Cuban religions of African descent in a bookshop in São Paulo in 2001, she found by way of

26 Machado: Posfácio, pp. 275–280.

27 Amado: *Mar Morto*, p. 265.

28 The Macumba ceremony is the focus of the seventh chapter, entitled "Macumba". When the ceremony begins, the leader priestess tia Ciata prays to three water goddesses: "Ôh Iemanjá! Anambuna! E Oxum! Tres Mães-d'agua". Mário de Andrade: *Macunaíma o herói sem nenhum caráter.* Santiago de Chile: Editorial Universidad de Chile 1997 [1926], p. 58.

29 Ana Maria Gonçalves: *Um defeito de cor. Romance.* Rio de Janeiro: Editora Record 2011 [2006].

30 The novel was written almost simultaneously with another more autobiographical text, *Ao lado e à margem do que sentes por mim* (edição do autor 2002), of which the first chapter is published in the Internet: http://www.manuelcarvalho.8m.com/ana4.html (accessed January 27, 2014).

serendipity another book instead, *Bahia de todos os Santos. Guia de ruas e mistérios*.[31] It contains the cultural history and description of Salvador by Jorge Amado, as a passionate declaration of love to his adored bay city, and is still re-edited today.

In *Um defeito de cor*, Salvador da Bahia is a very important place for the African woman Kehinde, whose life in the nineteenth century is told. She is enslaved and brought from Ouidah to Brazil and then, after 30 years, is returning from there to Africa. The title of the novel is mentioned only twice in the text in relationship with the Catholic religion. The first time, after having born her first son as a slave on a plantation on the island Itaparica near Salvador, Kehinde fears the possibility that through the mediation of the Catholic priest Father Notório her son—she feels him being African—will get dispensation from the "deficiency of color", a rule that forbids "colored" people to exercise an official function in religion, in the government, or in politics.[32] But then, several decades later, in discussion with a French Catholic priest in Lagos, Nigeria, Kehinde realizes that this "deficiency of color" is an obstacle not only in Brazil but even in Africa itself:

> I thought that only in Brazil did black people have to ask for exemption from the deficiency of color in order to be priests, but I saw that that was not true, it was also like that in Africa. Instead, in Africa the whites should be considered deficient, because it was our land and we were the majority. What I thought at that moment, but did not say, was that I felt myself much more human, perfect and triumphant than the priest. I do not have any deficiency and, for me, maybe to be black was and is a big quality, because if I would have been white, nothing would have forced me so hard to prove my capacities, life would not have rewarded so much effort with so much success.[33]

This fragment documents that *Um defeito de cor* is written in the first person. The lyrical subject is a fictional character, born in the Fon village

31 Jorge Amado: *Bahia de todos os Santos. Guia de ruas e mistérios*, illustrated by Carlos Bastos. Rio de Janeiro: Record 1977 [1944].

32 Gonçalves: *Um defeito*, p. 337.

33 "Eu achava que era só no Brasil que os pretos tinham que pedir dispensa do defeito do cor para serem padres, mas vi que não que em África também era assim. Aliás, em África, defeituosos deviam ser os brancos, já que aquela era a nossa terra e éramos em maior número. O que pensei naquela hora, mas não disse, foi que me sentia muito mais gente, muito mais perfeita e vencedora que o padre. Não tenho defeito algum e, talvez para mim, ser preta foi e é uma grande qualidade, pois se fosse branca não teria me esforçado tanto para provar do que sou capaz, a vida não teria exigido tanto esforço e recompensado com tanto êxito." Ibid., p. 893.

Savalu on the West Coast of Africa in today's Benin. When Kehinde is seven years old, she moves with her family to Ouidah, where they are captured and enslaved in 1817. She is brought to Itaparica and sold to a landowning family, with whom she later moves to the city of Salvador. This urban environment gives Kehinde a possibility to earn money and she becomes quite a successful "escrava da rua", a street vendor who gives part of her earnings to her mistress and subsequently learns the details of commercial enterprise, first making English cookies and puddings in the Historical Center of Salvador, and then, after being forced to leave Salvador for political reasons, settling in Rio de Janeiro, where she starts a business in articles from Bahia, such as tobacco and clothes. After several journeys within Brazil, Kehinde returns to Africa, first to Ouidah in 1847 and then, in 1863, to Lagos. At the end of her life, after 52 years in Africa, Kehinde again embarks in Lagos to return to Brazil.

The book ends here on the boat and this ending makes this transatlantic passage a major point. The first passage on the slave ship from Ouidah to Brazil was made under dramatic circumstances—she loses her twin sister and her grandmother; the second voyage, back to Africa, is on board the *Sunset*, an English ship, utterly comfortable in comparison with the first trip. Now Kehinde represents the typical *returnee* in the nineteenth century, the slave who managed to be manumitted and returned from Brazil to Africa. Historians have studied these returnees extensively, and there is a long list of literature on this phenomenon.[34] The most frequently quoted personality in this respect is Francisco Félix de Souza (1754–1849), a Brazilian-born slave, who became a famous slave trader himself on the West African coast. In fact, Souza appears as a character in Gonçalves' novel, being indirectly responsible for the capture of Kehinde and her family. But Kehinde does not want revenge on him. After coming back from Brazil to Ouidah thirty years later, she even seeks his protection for her business, first for selling arms. But then, later on, she starts specializing in designing houses in Brazilian style in Ouidah. After moving to Lagos, Kehinde changes her

34 Silke Strickrodt: The Brazilian Diaspora to West Africa in the Nineteenth Century. In: Ineke Phaf-Rheinberger / Tiago de Oliveira Pinto (eds): *AfricAmericas. Itineraries, Dialogues, and Sounds*. Madrid / Frankfurt am Main: Iberoamericana-Vervuert 2008, pp. 36–68. See also Christine Meisner: The Recovery of an Image: A Video Tale. In: Ibid., pp. 69–76.

African name to Luísa and becomes a well-to-do woman. She is even involved in the construction of the cathedral, which explains her conversation with the Catholic priest about the deficiency of colour. Luísa is proud of the quality of life in her Brazilian Quarter:

> We have many shops with nicely imported commodities, besides being expensive, at the Campos Square, Campbell Street, Kakawa Street and Bangbosh Street on Tibunu Square, where the Rocha Restaurant is situated, a simple restaurant and a hotel, but very well administrated.[35]

António Olinto clearly inspired Gonçalves to choose a woman as a main character. She mentions his novel *A casa da água* (Water House) in the reading list at the end of her book.[36] Olinto had "discovered" this Brazilian Quarter working at the Brazilian Embassy in Lagos. The Nigerian critic Niyi Afolabi is also familiar with this Quarter and writes about the impression Olinto's book made on him:

> Because I was born and raised in the bubbling heart of Lagos, Nigeria, Olinto's narratives, especially *A casa da água*, bring back memories of familiar settings. I can hear echoes of hustling street traders, the nocturnal pleasantries of celebrants barricading whole blocks for extravagant open-air parties, the playground in Campos Square (*Praça campos*) that serves as a point of encounter for students, truants, and adults who have come together to play soccer. Most significant is the Brazilian Quarter sandwiched between Kakawa, Broad, and Bamgbose Streets, as well as the relics of Brazilian architecture on Bamgbose, Simpson, and Oshodi Streets where the Iron Gate and a few other structures seem to transpose the northeast of Brazil, especially Bahia, into a Nigerian reality. Olinto's narratives depict an unbroken connection in this sense – a linkage that is easily made, a kinship that can neither be denied nor minimized.[37]

4. Gender, Naming and Spirituality

Gonçalves refers in *Um defeito de cor* to this continuous exchange of goods and people that linked the Brazilian Quarter in Lagos and in Ouidah to Salvador da Bahia. Here, in Africa, the *returnees* represent modernity due to their international contacts and commercial activities.

35 "Temos muitas lojas com artigos bonitos e importados, além de caros, ao longo da Campos Square, da Campbell Street, da Kakawa Street, da Bangbosh Street e da Tibunu Square, onde também fica o Restaurant do Rocha, hotel e restaurante simples, mas muito bem administrados." Gonçalves: *Um defeito*, p. 909.

36 Antônio Olinto: *Water House*, trans. by Dorothy Heapy. New York, NY: Carroll & Graf Publishers 1985; id.: *A Casa da Água*. São Paulo: Círculo do Livro 1975.

37 Niyi Afolabi: *Afro-Brazilians Cultural Production in a Racial Democracy*. Rochester: University of Rochester Press 2009, p. 108.

Kehinde's life is described in chronological order in ten chapters, divided by entitled subchapters and headed by African proverbs in Portuguese, which point to the conversational style of the narrative. Mineke Schipper has focused on proverbs to highlight their role in judging gender situations all over the world.[38] Gender relations determine Kehinde's personal life indeed. She has traumatic experiences with violence as a child in Africa, when she observes her mother being violated and killed by African warriors. Thereafter, in Brazil, she experiences violence herself. As a young slave girl, Kehinde is abused by the plantation owner and bears a son. Then, when living with the Portuguese Alberto in Salvador at the beginning of the 1830s, she gets her second son aborted, whose father Francisco is a slave. With this Alberto she conceives a third son, who is sold into slavery by his own father. Kehinde desperately starts searching for him in Rio de Janeiro and São Paulo and, after leaving for Africa, she feels constantly guilty for having left this "lost" son behind.

Gonçalves' novel contains an incredible amount of detailed information, not only on the topography of Salvador da Bahia, Rio de Janeiro, the Reconcavo and Lagos, but also on different religions and persons. In this manner, the reader of the book becomes familiar with *orishas* and *voduns*, as well as with words and customs from different parts of Africa and Brazil. Kehinde is far from being a heroic person; she always questions herself, her decisions and her way of being. The precariousness of her daily circumstances is portrayed not only in the context of oppression, violence, poverty and politics, but especially in her relationships with male partners and friends. Only her last two children are born in liberty in Africa. They are twins, a girl and a boy, from a "mulatto" father from Sierra Leone she became acquainted with on board the *Sunset* heading for Ouidah in 1847.

Kehinde's strategy for naming herself and her children reveals a specific logic explaining the difference between Africa and Brazil. She lost her twin sister on the slave ship to Brazil and has a traditional Yoruba name, given to the last-born of female twins. In Brazil, Kehinde receives a "white" name, her Christian name, Luísa Gama, but she continues calling herself Kehinde. Therefore, in addition to their Christian

38 Mineke Schippers: *Never Marry a Woman with Big Feet. Proverbs from Around the World.* New Haven / London: Yale University Press 2004.

names, she gives each of her Brazil-born sons an African name: her first-born Banjoko, who died young, and Omotunde Adeleke Dambira, the "lost" son. Yet, her Africa-born twins only receive "white" names, Maria Clara and João, whereas she herself starts bearing the name Luísa Andrade da Silva, Dona Luísa, or also Senhá Luísa. In Africa she enjoys the Brazilian sound of the Portuguese language, and naming characterizes her social status. Whereas in Brazil Kehinde sticks to her African identity, after her return to Africa she adopts a Brazilian identity, since as a "Senhá" her status is as high as that of the landowning family she served as a slave in Brazil.

Notwithstanding the various changes in her life, Kehinde's religious orientation gives her self-esteem and security. After arriving in Brazil and having lost her family members to a premature death, she immediately reconnects with her Dahomey background. As a slave girl on Itaparica, Kehinde meets an old Dahomey-woman, Nega Florinda, who was captured more than sixty years earlier and has been free for the last thirty years. This woman instructs Kehinde about religious differences between Brazil and Africa:

> In Brazil, the cult of the orishas was strong, notwithstanding the fact that the voduns possessed great power. She also said that I could use the orishas to venerate some voduns because in Bahia, Mawu, Khebioso, Legba, Anyi-ewo, Loko, Hoho, Saponan and Wu were worshipped as well as Olorum, Xango, Elegba, Ochum, Iroco, Ibejis, Xapona and Olokum. In Bahia, the orishas had already taken the heads of black people and they had been venerated for a long time, practiced by almost all Africans who, for many, many years, went to those lands. Our voduns never had enough strength to gain a bit of space or attention, and for them, there was a place not far from here. But about that, at this time, she could not speak any further.[39]

The difference between *voduns* (Dahomey) and *orishas* (Yoruba) remains a point throughout Gonçalves' novel, but the author does

39 "No Brasil, o culto aos orixás era forte demais até para o grande poder que os voduns possuíam. Ela também disse que eu poderia me valer dos orixás para cultuar alguns voduns, porque, na Bahia, Mawu, Khebiosô, Legba, Anyi-ewo, Loko, Hoho, Saponan e Wu eram cultuados como Olorum, Xangô, Elegbá, Oxum, Iroco, Ibêjis, Xaponã e Olokum. Na Bahia, os orixás já tinham tomado conta das cabeças dos pretos e o culto deles vinha de muito tempo, praticado por quase todos os africanos que, por muitos e muitos anos, iam parar naquelas terras. Nossos voduns nunca teriam força para ganhar um pouco de espaço ou atenção, e para eles estava destinado um lugar não muito longe dali, do qual, por enquanto, ela nada podia falar." Gonçalves: *Um defeito*, pp. 83–84.

not establish definitive boundaries between the two religions. Since childhood, Kehinde knows that "a god can be called with various names".[40] Kehinde is a daughter of Ochum, the sweet-water goddess, who is closely connected to Yemayá. Therefore, the encounter with the ocean on Kehinde's first trip as a child on the river from Savalu to Ouidah—the "ocean city"—gives her the impression of something already familiar to her: "I thought that the sea had the colour of the Yemanjá's cloth that my grandmother had in Savalu, only more brilliant and smoother".[41]

Nega Florinda takes Kehinde from Itaparica for a first visit to Salvador in order to meet Queen Agongela from Dahomey, who was also forced to go into the diaspora. Kehinde receives a wooden Oxum statue from her as a present. This statue accompanies Kehinde throughout life and, when she has her own house, she constructs a special room close to the sea for it, "a small place, beautiful and full of luxury things to satisfy Oxum's vanity, where I constructed an altar for the image".[42] This statue turns out to hide a treasure, with which she is able to buy her freedom, and it also travels back with her to Africa. There, in Ouidah and Lagos, Kehinde again installs a special room for worshipping such animated objects in her houses. The importance of their presence is emphasized when she lives through a particular dramatic moment, her daughter almost dying in childbirth. Kehinde prays to her *voduns* and *orishas*, completely confident that everything will have a happy end. This is the case and the episode testifies that, even when being a well-to-do person in a Christian environment and with a Christian name in Africa, the *orishas* and *voduns* still give her a feeling of stability.

5. The Hausa Rebellion and Other Intertextualities

In her prologue, Gonçalves tells that she was very much impressed to read about the Hausa Rebellion in Salvador in January 1835. Therefore, she not only concentrates on *orishas* and *voduns* from the Nigeria-Benin regions, but *Um defeito de cor* seems to plea to bring more diversity into the African heritage in Brazil. Her fascination with the Hausa slaves

40 "[U]m deus pode ser chamado por vários nomes." Gonçalves: *Um defeito*, p. 45.

41 "Eu achei que o mar era da cor do pano de Iemanjá que a minha avó tinha em Salavu, só que mais brilhante e mais macio." Ibid., p. 30.

42 "Um lugar pequeno, bonito e cheio de luxos para satisfazer a vaidade de Oxum, onde montei altar para a imagem." Ibid., p. 357.

from the Malê region in today's northern Nigeria derives from her admiration for their intellectual capacities. The long descriptions of Muslim life in Salvador—Kehinde temporarily even lives with them in the same house—are unique for Brazilian fiction set in the times of slavery. Historical research resulted in the publication of important volumes on this topic,[43] but fiction writers have been less interested in this Muslim dimension. Even the famous novel *Viva o povo brasileiro* by João Ubaldo Ribeiro, partly situated in Bahia in the era when Brazil gained independence around 1822, does not mention their existence.[44]

For Kehinde, in contrast, it is extremely important that these Hausa slaves can read and write and also possess books. They definitely have a higher cultural level than slaves from other African regions in this respect. Kehinde herself learns to read and write from Fatumi, the Hausa teacher in the Big House on Itaparica with whom she remains friends while living in Salvador. Through him she becomes involved in the Malê rebellion in January 1835, the *jihad* against the whites. Therefore, although the author mentions in her prologue January 2001 as the date on which she experienced her "serendipity", she might also have been influenced some months later by September 11, 2001. Her novel is constantly dealing with history and fiction. Kehinde's Christian name "Luísa Gama" is not accidental but alludes to the mother of a well-known Brazilian writer of African descent in the nineteenth century, Luiz Gama, who had the same experience of having been sold into slavery by his own Portuguese father.[45]

The anti-slavery discourse pervades Kehinde's life while she is in Brazil. In Africa, however, this becomes less important because Luísa's

43 João José Reis: *Rebelião escrava no Brasil. A história do levante dos Malês em 1835*, ed. revisada e ampliada. São Paulo: Companhia das Letras 2003. This book is included in the reading list at the end of Gonçalves' novel.

44 João Ubaldo Ribeiro: *Viva o povo brasileiro.* Rio de Janeiro: Nova Fronteira 1984. This book is included in the reading list at the end of Gonçalves' novel.

45 Luiz Gonzaga Pinto da Gama was born in Salvador in 1830 and died in São Paulo in 1882. His mother was a *nago*, born in Africa, and his father a representative of one of the most important families of Portuguese descent. See Maria Consuelo Cunha Campos: Luiz Gama. In: Eduardo de Assis Duarte (ed.): *Literatura e Afrodescendência no Brasil: antologia crítica*, vol. 1: Precursores. Belo Horizonte: Editora UFMG 2011, pp. 127–142.

business partners are all involved in this trade. She is aware of this change in her orientation:

> Sometimes I remained a bit embarrassed because of my involvement with slave traders, but then I forgot it, because that was not my problem anymore. I couldn't solve it even if I tried, and I really could not have too many scruples after providing arms to King Guezo, knowing that they would be used in wars for slaves, almost all of them to be sent to Brazil.[46]

Kehinde's fictional life is not only embedded in the history of Brazil, but also of Africa, both constantly affected by the interventions of the English in the slave trade. Yet, although not mentioned in Gonçalves' list of references included at the end of her novel, also Carolina Maria de Jesus and her *Child of the Dark* (Quarto de despejo) come to mind when reading her manuscript. Carolina, a black inhabitant of a favela in São Paulo in the 1950s, became famous overnight when her diary was published; in it she describes her daily struggle for her own and her three children's survival, listing exactly what she earns and spends. She often does not have enough money for the most basic needs, such as medical care and school material. Her descriptions were so appealing to a 1960s reading audience that her diary was immediately translated into many languages and is still often used as an obligatory school text in Brazil today.[47] Entries like the following are frequently found:

> I had bought an egg and lard for 15 cruzeiros at Seu Eduardo. I baked the egg to look whether my dizziness would stop. It stopped. I understood that it was because of weakness. The doctor ordered me to eat oil but I cannot afford it.[48]

Carolina's text had a deep impact on Brazilian authors of African descent. Conceição Evaristo states that it motivated female black favela inhabitants to learn to read and write because Carolina could

46 "Ás vezes eu ficava um pouco constrangida por me relacionar com mercadores de escravos, mas logo esquecia, já que aquele não era problema meu. Eu não conseguiria resolvê-lo mesmo se quisesse, e também não poderia ficar com muitos escrúpulos depois de fornecer armas para o rei Guezo, sabendo que seriam usadas em guerras que fariam escravos, quase todos mandados para o Brasil." Gonçalves: *Um defeito*, p. 771.

47 Carolina Maria de Jesus: *Quarto de despejo*. 8th ed. São Paulo: "Livr. Francisco Alves" / Paulo de Azevedo 1960. English translation: Carolina Maria de Jesus: *Child of the Dark: the Diary of Carolina Maria de Jesus*, trans. by David St. Clair. New York: New American Library 2003.

48 "Eu havia comprado um ovo e 15 cruzeiros de banha no Seu Eduardo. Eu fritei o ovo para ver se parava as náuseas. Parou. Percebi que era fraquesa. O médico mandou-me comer oleo mas eu não posso comprar." Jesus: *Quarto*, p. 48.

do so, though enduring the same experiences of hunger and poverty as everybody else.[49] Carolina is not only eager for education, but also loves to talk to educated people; she shares these characteristics with Kehinde, Gonçalves' fictional character, a street vendor and thereafter owner of a bakery shop continuously reflecting on the prices and numbers of cookies.

Although *Um defeito de cor* is a novel, we find in it many traces of other genres, such as diaries, autobiographies, historical works and anthropological descriptions of religion. Kehinde's / Luísa's life story is all the more important, because, with the exception of Cuba, [50] Latin America does not possess written sources from (former) slaves in the nineteenth century, as is the case in the USA. In this sense, the letter written by Luiz Gama at the end of his life[51] might have been a major inspiration for suggesting in *Um defeito de cor* that Kehinde writes her life story, starting at the turn of the nineteenth century, three days before embarking on the ship in Lagos, and dictating to her friend, who writes it down on paper. Kehinde is familiar with writing. Being in business in Africa, she maintains an extended correspondence with the daughter of her former slave-owning family, with whom she grew up and shared many experiences, as well as with her business partners in Brazil. She mentions books like the Bible, the *Lusíadas* by Luis de Camões and the *Cartas* (1669) by Sóror Mariana Alcoforado, as well as sermons by Padre António Vieira. The most important reference is to José Saramago's *Memorial do Convento* (1982). Gonçalves underlines this intertextuality, placing a footnote to this reference on the page on which she speaks about Kwanza. When living in Brazil, this man brings her to a Kianda celebration to worship the sea goddess of the Kimbundu traditions in Angola, and Kehinde does not know what to admire more: the yearly celebrations for Brazilian independence or the Kianda ceremonies. Kwanza is an old man living in the city of

49 Conceição Evaristo. In: Eduardo de Assis Duarte / Maria Nazareth Soares Fonseca (eds): *Literatura e Afrodescendência no Brasil: Antologia crítica*, vol. 4: História, teoria, polêmica. Belo Horizonte: Editora UFMG 2011, p. 105.

50 Juan Francisco Manzano: *Autobiografía del esclavo poeta y otros escritos*, ed., intr. and notes by William Luis. Madrid / Frankfurt am Main: Iberoamericana-Vervuert 2007.

51 Luiz Gama wrote a long letter about his life to Lúcio de Mendonça two years before his death on 25 July 1880, which was published in the *Almanaque literário de São Paulo para o ano de 1881*. See Campos: Luiz Gama, pp. 131–135.

Cachoeira, not far from Salvador. His father, who died in 1750, was from Angola and named his son after the main river in his home country, the Kwanza. This son is more than ninety years old when he starts telling Kehinde that his father was the slave of Father Bartolomeu Lourenço de Gusmão (1685–1724), a historical personality, regulator of the water supply system in the Bahia region and inventor of the first hot air balloon, the Passarola. Gusmão and his slave travelled to Portugal and then to Spain, where the Father died. Kwanza's father returned to Brazil with all his documents concerning the construction of the flying machine. This story impresses Kehinde so deeply that she has a dream in which Kwanza tells her that because of his name he always identifies with being a river; all rivers stream into the sea and the sea brings people everywhere and might also bring Kwanza back to the land his father originally left. Kehinde dreams how Kwanza starts traveling to the land of his father:

> And the most interesting thing was that he did not use a boat but instead those sheets of paper that he had guarded so carefully. He placed one of them on the water and then this sheet unfolded, forming a lengthy and delicate tapestry fixed on the sea as if it was land, sufficiently strong to bear Kuanza's body. When he came to the end of one sheet, he put the next one on the water, which also unfolded so far that you didn't see the end, bringing him further, all the way to Africa. Each sheet took the format of a small and quiet river, just like he thought the Kwanza should be.[52]

In this context, Kehinde philosophizes about the lack of meaning of writing for African people.[53] Anyhow, after learning that her "lost" son has become an educated man, she finally does not hesitate to write down in the most detailed way her own findings. Without a doubt, the happiest period in her life is as Dona Luísa Andrade da Silva in Lagos—"I was happy in Lagos"[54]—whose archipelago-like structure

52 "E o mais interessante era que não usava embarcação nenhuma, mas sim aquelas folhas de papel que guardava com tanto cuidado. Colocava uma delas sobre as águas e logo ela se espichava, formando um tapete comprido e fino assentado sobre o mar como se fosse sobre a terra, suficientemente firme para sustentar o corpo do Kuanza. Ao chegar ao fim de uma folha ele deitava outra, que também se espichava até se perder de vista, levando-o mais adiante, e assim até a África. Cada folha tomava o formato de um rio estreito e tranqüilo, como ele achava que o Kuanza devia ser." Gonçalves: *Um defeito*, p. 617.

53 "[O]s africanos não gostam de pôr histórias no papel, o branco é que gosta." Ibid., p. 617.

54 "[F]ui feliz em Lagos." Ibid., p. 887.

is emphasized. The numerous islands and the Brazilian Quarter have high cultural standards and, although Luísa / Kehinde almost never leaves her house because she is old and blind, she knows everything about Lagos' geography and inhabitants. The "Brazilians" are clearly at the top of local society: "for us, the Brazilians, it is work, intelligence and initiative that made Africa grow".[55]

6. Luanda

In contrast to Brazil, the role of the sea stands out in the literature from Angola. For this topic, the work of the Brazilian Carmen Secco is particularly relevant. She edited the *Antologia do Mar na Poesia Africana de Língua Portuguesa do Século XX. Angola* (Anthology of the Sea in African Poetry of the Portuguese Language in the Twentieth Century)[56], in whose introduction she explains that when she started to teach African literatures at the Federal University of Niterói in the State of Rio de Janeiro in 1993, appropriate instructional material was lacking. So that same year Secco started a research project on the sea, myths and memory in 20th-century African poetry, taking the following central thesis as her starting point:

> When observing the history of Portuguese colonisation in Africa and America, we verify that the details of the cultural memory of the two continents were blocked by the route of the conquest, and that this caused a "paralysis of memory", an amnesia regarding traditions. Extinguished were the African and American traces and marks, the impact and links of cultural identity were brought to a standstill. The orality of the natives was silenced, and the sea, inhabited by myths and African gods, became "Portuguese", the domain of important persons and heroes in which the white invaders took pride.[57]

Secco formulates some major points about important stages of history: 1) the Portuguese sea as the subject of the colonial ideology of

55 "[P]ara nós, os brasileiros, o trabalho, a inteligência, e a iniciativa que faziam a África crescer." Ibid., p. 907.

56 Carmen Lúcia Tindó Ribeiro Secco (ed.): *Antologia do Mar na Poesia Africana de Língua Portuguesa do Século XX. Angola*, vol. I. Luanda: Kilombelombe 2000.

57 "Observando a história da colonização portuguesa na África e na América, verificamos que as trilhas da memória cultural desses dois continentes foram bloqueadas pela rota da conquista, o que acarretou uma "paralisia do lembrar", uma amnésia das tradições. Apagados os rastos e as marcas africanas e americanas, imobilizaram-se os percursos e os laços de identidade cultural. A oralidade dos nativos foi silenciada e o mar, morada de mitos e deuses africanos, tornou-se "português", domínio de vultos e heróis ufanados pelos invasores brancos." Secco: *Antologia do Mar*, p. 39.

the caravels and the Portuguese adventurers, an idyllic scenery with exotic connotations; 2) then, in the period of anti-colonial struggle, the sea was seen from a revolutionary perspective, as a symbol of freedom as well as the holder of the tragic memory of the slave trade; 3) in the period of national reconstruction, the sea became the immense reservoir of the unconscious *afro,* silenced by conquest and colonization; and 4), in the 1980s and 1990s, a "new sea" (mar novo) emerged connected to the female image, with eroticism and lyrical sensuality.

Most of the older poets in Secco's anthology can be found in Manuel Ferreira's anthology, published in Lisbon,[58] but Secco also presents a list of authors who started publishing afterwards, such as João Melo, António Gonçalves and José Luís Mendonça. Her book includes texts from 42 authors, who published from 1849 until 1995, of which the last poem, "Noite (Ki)anda" ((Ki)anda's Night)[59] by John Bella is particularly relevant to this essay. Bella's poem plays alternatingly with Portuguese and Kimbundu words, such as *muzonge* (fish plate), *mulemba* (wild fig tree), *axilwanda* (Luanda people), *misanga* (knotted jewellery), *besangana* (a respected woman dressed in typical clothes) and *xinguilar* (falling in trance). On top of that, Bella also creates tension between Kianda, the ocean goddess, and Kalunga, the association of the ocean god with the slave trade and death, who puts a spell on the *kalumba* (young girl).

Bella's double orientation toward the identification with the ocean is revealing; it marks the transition from Kalunga to Kianda[60] in the effort of overcoming the "blocked" memory of the past. Kalunga is the personification of this hidden trauma, and it was Agostinho Neto who first referred to him in his short story *Náusea* (nausea, 1952), interpreted by Ana Cristina Moura Alves de Morães as the link with fate and disaster.[61] Tania Celestino de Macêdo reproduces a quote from Neto: "And the sea, is always Kalunga. Death. The sea brought grandfather to

58 Manuel Ferreira (ed.): *No reino de Calibán. Antologia panorâmica da poesia africana de expressão portuguesa*, vol. II: Angola, São Tomé e Príncipe. Lisbon: Seara Nova 1976.

59 John Bella: "Noite (Ki)anda". In: Secco: *Antologia do Mar*, pp. 245–246, here p. 246.

60 Kianda is also written as Quianda or Kyanda. I will use Kianda in this essay, except in quotations.

61 Ana Cristina Moura Alves de Morães: O Mar sob o Ponto de Visto de Agostinho Neto no Conto *Náusea.* http://www.ueangola.com (accessed October 14, 2013).

other continents. Kalunga is slave work. The sea is the enemy".[62] The Brazilian critic Benjamin Abdala Júnior subscribes to these arguments. He compares Neto's short story with Sartre's *La nausée* (1938), emphasizing the difference; Neto's *Náusea* focuses on the remembrance of the horrors of colonial times.[63]

Tania Macêdo observes in her book *Luanda, cidade e literatura* that, after Agostinho Neto, authors in Angola abandoned the theme of the sea. Jofre Rocha, Luandino Vieira and António Jacinto concentrated on the poor neighborhoods—*musseques*—of Luanda in their prose.[64] However, when we read Neto's short story, we should not forget that the main character—an older man—lives in a *musseque*, although he was born on the Ilha, the island in the Bay of Luanda, from where he migrated to the city.[65] Neto's poetry frequently portrays the people from the *musseques*, for example in his poem "Kalumba", in which the daughter of a *soba* (local chief) who migrated with his family from Lunda in the interior to the neighborhood Sambinzanga in Luanda observes the sea: "But her confused eyes / discovered in the city / a different world / which confined her soul / to the ships that brought / people from Congo / over the sea / Kalunga! Death".[66]

We have found several speculations about the etymology of Kalunga. Marie-Louise Bastin writes in *Arte Decorativa Cokwe* (2010) that Kalunga means God in a traditional iconography drawn in the sand,

62 E o mar, é sempre Kalunga. A Morte. O mar tinha levado o avô para outros continentes. O trabalho escravo é Kalunga. O inimigo é o mar. In: Tânia Celestino de Macedo: Visões do mar na literatura angolana contemporanea. http://www.fflch.usp.br/dlcv (accessed October 14, 2013).

63 Benjamin Abdala Júnior: Images of a Colonial Kalunga. In: *Research in African Literatures* 36,1 (2007), pp. 68–74.

64 Tania Macêdo: *Luanda, cidade e literatura*. São Paulo: Editora UNESP 2008, p. 119.

65 The Ilha in the Bay of Luanda has a special history. When the Portuguese arrived, it was the property of the King of Congo, who collected here the *zimbos*, the little shells used as money in his territory. The Portuguese, who had already established contact with this king before, were permitted to build their first church on the Ilha in 1571. The islanders called themselves the *axiluanda*, the people from the sea, and this became the name of the city of Luanda. Today the Ilha is connected to the city and motivated Maria Celestina Fernandez to write the novel *A muxiluanda* (Luanda: Chá de Caxinde 2008) on its customs and habits in past and present.

66 "Mas seus olhos confusos / descobriram na cidade / un mondo diferente / onde a sua alma era aferrolhada / nos navios que levaram do Congo / os homens sobre o mar / Kalunga! Morte!" Agostinho Neto: Kalumba. In: Id.: *Sagrada Esperança, Renúncia Impossível, Amanhecer*. Luanda: UEA 2009, p. 150.

representing the origin of creation in the Chokwe cosmology in Lunda, in the interior of Angola.[67] Kalunga probably migrated from the interior to the coast,[68] in contrast to Kianda, whose origin seems to have been in this coastal Kimbundu region. The anthropologist Virgílio Coelho describes how Kianda's name indicates the spiritual power of "seres fantásticos" (phantastic beings) or a "genius of nature" (génio da natureza).[69] He claims that the word *kiyàndà* historically developed in North-West Angola and implied the association with the spiritual aura of the sea, the rivers or the humid spots in the forest, as well as the capacity to live in as well as outside of the water. *Kiyàndà* would manifest herself either as a man or a woman. Under European influence, these local associations gradually became more unified in the image of the mermaid, half woman and half fish. Coelho mentions that as early as 1687 Padre Cavazzi depicted a siren (mulher-peixe) in the Kwanza River in his *Historical Description of the three Kingdoms of Congo, Angola and Matamba*. In Coelho's opinion, this unilateral association of *kiyàndà* with the mermaid is misleading, although even by the Angolans themselves have adopted it. He therefore emphasizes the importance of reflecting on its original meaning and on the values of the Kimbundu mind-set.

7. Slavery in History

It is obvious that, today, it is not Kalunga but Kianda who appears to be related to the memory of slavery in contemporary poetry and prose in Angola. She seems to be everywhere. Titles like *Na boca árida da kyanda* (in the dry mouth of Kianda) of a poetry volume by Chó do Guri, and *O desejo de Kianda* (Kianda's desire), a novel by Pepetela, are only some examples among many others.[70] Macêdo even characterizes

67 Marie-Louise Bastin: *Arte Decorativa Cokwe*. Coimbra: Muséu Antropológico da Universidade de Coimbra 2010, p. 36.

68 Adriano Parreira: *Dicionário de etnologia angolana*. Porto: Plural Editores 2013, p. 292.

69 Virgílio Coelho: "Imagens, símbolos e representações, 'Quiandas, Quitutas, Sereias': Imaginários locais, identidades regionais e alteridades. Reflexões sobre o quotidiano urbano luandense na publicidade e no universo do marketing." In: Id.: *Os Túmúndòngò, os «Génios» da Natureza e o Kílàmbà. Estudos sobre a sociedade e a cultura Kímbùndù*. Luanda: Kilombelombe 2010, pp. 142–194, here pp. 228, 226, 243.

70 Chó do Guri: *Na boca árida da kyanda*. Luanda: Kilombelombe 2007; Pepetela: *O desejo de Kianda*. Lisbon: Dom Quijote 1995.

Kyanda as the "most illustrious resident"[71] of this city. We have also found several narratives, in which Kianda becomes related to slavery in historical novels or in narratives about contemporary Luanda.

Pepetela, one of Angola's best-known authors, has written *A Gloriosa Família* (the glorious family) about city life in the oldest part of town in Early Modern History.[72] The plot is set in the period from February 1641 to October 1647, the time of the Dutch occupation of Luanda and of Recife in Brazil, and portrays this transatlantic connection, from the very beginning, as the representation of the slave trade.[73] The main character is the Ngola slave of a Flemish merchant, Baltasar Van Dum, who arrived at this "Bay of Dreams" in October 1616, seeking his fortune overseas. He is involved in the slave-business, of course, and marries the daughter of an important *soba* having eight children with his official wife, as well as several others with slave women on his territory. We learn about the details of political and personal daily life from the perspective of Van Dum's personal slave, his "shadow" so to speak, who does not have a name and is born mute. His mother is a court lady of Queen Nzinga in the interior who had an affair with an Italian friar and gave birth to this son.

With this slave, Pepetela does not introduce a completely new type of character to the literature of Angola. The slave condition has been a trope since the novel *Nga Muturi* (ma'am widow), written by Alfredo Troni,[74] but the character in the latter book is not embedded in such a realistic historical background as in Pepetela's novel. The house of Van Dum is situated on a hill with a view of the bay, the harbor, the Ilha and the city. Through the marriages of its children, the Van Dum family is linked to Ngola, Congolese, Portuguese and Dutch people and uses those contacts to negotiate its interests with the interior and the exterior. The view from van Dum's house is repeatedly mentioned:

> From there you saw the whole Ilha, from the point closing the entrance to the harbour to the bar Corimba in the south. From there you saw the city and the hill

71 "[M]oradora mais ilustre." Macêdo: *Luanda, cidade*, p. 137.

72 Pepetela: *A Gloriosa Família*. Lisbon: Dom Quixote 1997.

73 Ineke Phaf-Rheinberger: *The Air of Liberty. Narratives of the South Atlantic Past.* Amsterdam / New York: Rodopi 2008, pp. 3–45.

74 Alfredo Troni: *Nga Muturi. Cenas de Luanda*, preface by M. António. Lisbon: Edições 70 1973. This novel was first published as a serial in the *Diário da Manhã* in Lisbon in 1882.

> of S. Paulo and the hill of Samba. And back to the east, you saw the way to the interior, to the kingdom that Ngola Kiluanje unified, the fatherland of the Ngola, my fatherland.[75]

This topographical precision corresponds with the description of places such as the Kinaxixi Lagoon, between Van Dum's house and Old Luanda, bordered by wild cotton and baobab trees inhabited by spirits, and this environment obviously functions as a symbol of some generally known taboos. When one of Van Dum's sons, Hermenegildo, discusses near Kinaxixi with the Catholic Father Tavares, a Portuguese, this priest approaches him indecently, which causes quite a complex situation.[76] The importance of this lagoon is additionally emphasized when one of Van Dum's daughters, Rosário, has an affair with the slave Thor, the son of a *soba* who was captured when hunting. The fact that his daughter has an affair with a slave is so unheard of for Baltasar that he decides to set an example. Thor is sentenced to death and brought to Kinaxixi; he is killed and falls into the water. Immediately afterwards, the lyrical subject observes a spectacular scenery:

> Thor's blood, floating on the surface, was transformed into the round leaves of water lilies, and stalks grew from them with white flowers. White flowers like those in jars and which emitted a very strong perfume.[77]

This homage to Thor's death is remembered later on when the Ngola slave feels the desire to take revenge on his "rival", one of the directors of the Dutch West Indian Company in Luanda, Heynderick van Redinckhove. From the first page of *A Gloriosa Família*, the reader is informed that the slave is in love with Catarina, a daughter of Baltasar born out of wedlock. Because of Thor's death, however, the reader understands why he never approaches her: a slave can never marry a member of the Van Dum family. Therefore, when Van Redinckove makes love to Catarina, the lyrical subject of the story has a dream in which, passing the Kinaxixi Lagoon—where the wronged souls, the

75 "Dali se via a Ilha toda, desde a ponta fechando a entrada para o porto até à barra da Corimba no sul. Dali se via a cidade e o morro de S. Paulo e o morro da Samba. E para trás, no sentido do oriente, se via a entrada da terra, o reino que Ngola Kiluanje unificou, a pátria dos Ngola, a minha." Pepetela: *A Gloriosa Família*, p. 19.

76 Ibid., pp. 206–208.

77 "O sangue de Thor, boiando à superfície, se transformava em folhas redondas de nenúfares e delas cresciam hastes com flores brancas. Flores brancas como as dos jarros e que exalavam um perfume muito forte." Ibid., p. 247.

drowned without decent burial and the kiandas with an unpredictable temperament live—Redinckove is killed and falls into the water, mixing his blood with that of Thor. But in this case, this fusion does not generate white flowers because this is a "justified" death, in contrast to that of Thor.

Pepetela, of course, is a master at portraying Luanda with great topographical precision and his plot is dominated by regional and international complications. His colleague Arnaldo Santos concentrates in *A Casa Velha das Margens* (the old house at the margins)[78] on this same Kinaxixi Lagoon and its specific spirituality in another historical epoch, on the period immediately after the Berlin Conference of 1884–1885, although he does not mention this conference explicitly. In those years, the colonial government of Angola attracted *novos colonos* (new farmers) from Portugal, and this implied that the existing complicated networks of moral responsibilities and commercial contacts in Luanda and its hinterland were destroyed. Santos' novel intends to reconstruct those networks from the perspective of ca. 1889, when Mendonça, the son of a Portuguese father and an Angolan mother, a Kissama woman, returns home after having studied in Portugal. He finds everything changed. His parents are gone and the old farmhouse of the family in the interior is burnt down. On top of that, Emídio is the victim of a violent assault that almost leads to his drowning in the river. This violent reencounter with his homeland causes such a shock that he is not able to express himself, which handicap requires him to find a new language in which he will be able to connect with his former life.

Emídio is a typical representative of a *filhos-do-país*-family, which feels Angola as its home country. The first lesson he learns after his return is that "in this country, whatever it is, you never carry something yourself. At least, when you want to be somebody".[79] Luanda is a city of servants and "free" people who work for others. Although the slave trade was officially abolished in 1836 and slavery in 1878, it obviously continues influencing social relations in reality. The new migrants from Portugal, often "degradados" with a criminal past, are not interested

78 Arnaldo Santos: *A Casa Velha das margens*. 2nd ed. Luanda: Edições Chá de Caxinde 2010 [1999].

79 "[N]esta terra, carregar, seja o que for, é a última coisa que deve fazer […] Isso, se quizer ser alguém." Santos: *A Casa Velha*, p. 42.

in the details of the intimate relationships among the local population. Santos characterizes this change in the contrast between the memory of the old farmhouse in the interior, a center for communication with the *sobas* and the Portuguese administration, and the new blue house at Kinaxixi, nearby up- and downtown Old Luanda, a symbol of its loss. Kinaxixi's marginal position in relation to the city equally reflects Emídio's own gradual exclusion of public life. A mad poet lives nearby and enigmatic visions and appearances intervene in his life, as do stories about Kianda, the water spirit of Kinaxixi. His sister, a "parda desvalida" (invalid colored woman), suffers from the sleeping sickness, whereas towards the end of the novel many occurrences and situations seem to become unreal, indicating an increasingly dramatic solitude and exile situation. To escape this chaos, Kianda leaves Kinaxixi, taking refuge in the Atlantic Ocean.

Santos' effort to reconstruct this "lost" world seems to complement José Eduardo Agualusa's *Nação crioula* (creole nation),[80] whose plot is set in that same period, the nineteenth century. The title of the book—creole nation—indicates the name of the—presumably—last slave ship from Angola to Brazil, which sailed in 1876. Agualusa chooses the genre of the epistolary novel for this book, presenting a collection of 25 letters, all written by the Portuguese Fradique Mendes except the last one, which is authored by the former slave Ana Olímpia. She is the daughter of a Congolese king, who was imprisoned by the Portuguese. Therefore, his daughter was born as a slave but lives as if she were a free woman, having been sold and then married to a much older merchant from Salvador da Bahia, who takes care of her education. When he dies, his brother comes to Luanda to sell Ana Olímpia as a slave again, from which deplorable condition Fradique liberates her, and they escape on the *Nação crioula* to Brazil.

Agualusa opts for portraying a Portuguese with liberal ideas centering his novel on the international debates about slavery—the secret correspondence. Through the eyes of Fradique, when he arrives in Luanda in 1868, the term *filhos-do-país* is a euphemism for people of "mixed

80 José Eduardo Agualusa: *Nação crioula. A correspondência secreta de Fradique Mendes.* Lisbon: Dom Quixote 1997. The book is translated into English: José Eduardo Agualusa: *Creole*, trans. by Daniel Hahn. London: Arcadia Books 2002. Unfortunately, I did not have a copy at my disposal when writing this essay and include here my own translations.

blood and black people with shoes" who seem to be always conspiring in bars and never agree upon anything because of their severe problems with the local hierarchy.[81] He is not interested in having contacts with these *filhos-do-país*, who do not represent modernity, Fradique being a fictional character from the work of the most important Portuguese modern writer, José Maria Eça de Quieirós (1845–1900). Agualusa's book is full of explicit literary intertextualities. While traveling on the *Nação crioula*, one of the crew interprets on his guitar some verses from *O navio negreiro* (the slave ship), the classic poem of Brazilian literature by the Bahian poet Castro Alves.[82] And Ana Olímpia remembers verses from António Gonçalves Dias about his passionate attachment to his home country, Brazil.[83] Agualusa's style is highly ironic when he describes what "work" meant in those days in Luanda:

> Work – nobody works in Luanda except the slaves; and outside of the city work the so-called "uncivilized blacks". Therefore, for the inhabitant of Luanda, work means an inferior activity, practiced by wild people or prisoners. "This fellow comes from a working class family", you sometimes hear somebody saying with a lowered, venomous voice, at the shady table of a bar. It is a cruel insinuation, capable of destroying reputations, because it suggests that the person under observation only recently bought his first pair of shoes and that he probably descends from slaves.[84]

Shoes, obviously, were a status symbol in Luanda's society. The urban topography in which Ana Olímpia lives is reduced to up- and downtown Old Luanda, to the houses of Europeans and of people completely oriented toward Europe and Brazil. Ana is successful and owns slaves herself on her plantation in Paraíba in Brazil, where she lives for a couple of years before returning to Angola. It is significant that she writes the last letter in August 1900 in Luanda herself, as an independent merchant woman after Fradique's death. She summarizes her

81 Agualusa: *Nação Crioula*, p. 16.

82 Ibid., pp. 72–73.

83 Ibid., p. 158.

84 "Trabalhar ninguém trabalha em Luanda a não ser os escravos; e fora da cidade trabalham os, assim chamados, «pretos boçais». Trabalhar representa portanto para o Luandense uma actividade inferior, insalubre, praticada por selvagens e cativos. «Fulano vem de uma família trabalhadora», ouve-se dizer às vezes em voz baixa, venenosa, à mesa sombria de um café. É uma insinuação cruel, capaz de destruir reputações, pois sugere que o visado só há pouco tempo comprou o primeiro par de sapatos e que provavelmente descende de escravos." Ibid., p. 16 (my translation).

life from her own perspective and the letter reveals that now, in her country, Ana is considered a *brasileira*, a returnee, who learned her distinguished manners and habits from Brazil. Ana herself adds to that argument by writing that one of the most important moments in her life was when she discovered the library of her first husband, a Brazilian from Salvador da Bahia. It made her feel like traveling on a ship, seeing the stars and the books above her, the universe, the unknown, the world.[85]

8. The Contemporary Memory

In Manuel Rui's visionary and satirical text *Memória de mar*,[86] meanwhile, we find a writer who relates city life and slavery in Luanda to the contemporary political context. He conceives the start of a "new world" after 500 years of slavery, in honor of the independence of Angola. His four main characters—a sociologist, a major, a historian and the narrative subject, the writer—undertake three scientific expeditions to document this change. They first visit the Ilha on 13 May, the Day of the Patron of Portugal, the Virgin of Fatima. The Ilha, inhabited by friars, servants and donkeys, is isolated and peaceful, whereas a war is being waged on the continent. It becomes clear that the servants come from the African continent and the religious friars from Portugal, which explains their difference in status. During the procession in homage to the Virgin of Fatima, Satan himself appears and everyone flees to safety in another place. Rui displays many twists on the Catholic religion, so different from the scientific mind of the four who are engaged in the expedition. At the same time, just before Independence Day, a submarine with Portuguese military is waiting at the coast, whose crew is completely convinced of its final victory over the freedom fighters. But it turns out differently, as we learn in the last three chapters. After having studied life "under the surface of the sea", where they found the submarine sunk off the coast, full of books in Portuguese, except for a catechism written in Kimbundu from 1642, these four men involved in the expedition enter Luanda, some time after 1975. They carry the chains with them from the last slave, who succeeded in escaping from the sinking Portuguese submarine, and

85 Agualusa: *Nação Crioula*, p. 149.

86 Manuel Rui: *Memória de mar*. Lisbon: Edições 70 1980.

deliver these souvenirs to the Slavery Museum. They then see a film in the film theatre, which has changed its name, which turns out to be a documentary realized on the Ilha, now a vacation paradise for the pioneers in socialist Angola. Instead of the former statue of D. Junqueira, the freemason, they observe a huge statue of Kianda, used as a playground for children. This Kianda, the "mermaid" of the Kwanza River, used to move all the oceanic waters against the Virgin of Fatima before she disappeared into the ocean, but not after having swallowed the last priest, Father Superior.

The memory of slavery in contemporary Luanda comes back in *O último segredo* (the last secret)[87] by Tazuary Nkeita, situated in the twenty-first century. The main male character Tuluca Dibaia is of poor descent and first worked as a communication engineer for the army. After having married F. F. Yianda, Tuluca has become a successful businessman. This Yianda, the synonym for Kianda, has the same name as the city of Luanda, indicated by a kilometer sign at the edge of the Kwanza River: "Cidade da Kyanda, 72 km".[88] She symbolizes the problematic balancing act between the "anciens et modernes"; she is like a bond and indicates that, in times of modernization, care for others should not be neglected. This fictional character F. F. Yianda has multiple meanings: she personifies the relationship with Kimbundu tradition; she is a modern woman; she tells wonderful stories to her children and husband; she is very practical in everyday life; and she takes the utmost care of her bodily appearance, as does the average middle class female (and male) in Luanda today.

The remembrance of slavery is a thread running throughout the book. Tuluca Dibaia gives his first speech in the Slavery Museum on Queen Ana de Sousa, or better, "Ginga Mbandi", the famous queen of the Ndongo people of Angola in the seventeenth century. The Dibaia family, however, does not live in the "Americanized" southern part of Luanda, but resides in a neighbourhood that might be identified with the Coqueiro quarter, close to the commercial centre of Old Luanda, the Bay and the Ilha. They meet the North American professor Joanna Woods, who is extremely interested in knowing more about the Dibaia's family background, as part of her historical research on

87 Tazuary Nkeita: *O último segredo*. Luanda: UEA 2011, p. 25.
88 Ibid., p. 30.

her roots financed by the Investigation Center for Black Culture in New York. However, only at the end of the novel does the reader learn that the father of the Dibaia family, Tuluca, has a slave background. Yianda finally finds out, after 35 years of marriage, this family "secret", which says something about the power of this taboo in contemporary urban society in Angola. A forefather of Tuluca belonged to a rebellious group of escaped slaves who distributed subversive propaganda against the slave trade in a colonial house in the same part of town in which the Dibaia family now lives.

Pepetela, too, experiments with introducing the representatives of new prosperous groups in contemporary Luanda. His main character *Jaime Bunda* is the offspring of a well-known family, proud of its four-hundred-year-old last name. He might even be a descendant of Pepetela's Van Dum family in the seventeenth century, due to the fact that Bunda's last name is still known because of having owned a large number of slaves in the past.[89] This contemporary Jaime is a detective working for a state organization, the Secret Service, similar to James Bond. But in contrast to Ian Fleming's hero, Jaime is just starting with his first case.

Bunda's parents used to live in the Barrio Operário, the quarter of the legendary music band Ngola Rhythms in the 1960s and 1970s. But when the family could not pay the rent any more, they had to move to Sambizanga, the *musseque*, a much poorer environment situated next to Roque Santeiro Market, the "biggest market in the open air of Africa",[90] where Jaime Bunda spent many hours in his youth. After his father's death, Jaime's uncle invites him to live in his house in Vila Alice, a central place in town with stone houses. Notwithstanding this shift in status, Jaime remains close to his habits of the Sambizanga quarter and the Roque Santeiro Market, both next to the harbour. These topographies likely are of strategic importance for foreigners, such as Arabs and people from the Middle East, who migrate for business to Luanda and are viewed with some suspicion, not least because of their Muslim religion. They try their luck in booming Angola, and Mulika, a dancer from Algiers who came with a Lebanese businessman, describes their

89 Pepetela: *Jaime Bunda, agente secreto. Estórias de alguns mistérios.* Rio de Janeiro, São Paulo: Record 2003, p. 249.

90 "[O] maior mercado ao ar livre da África." Ibid., p. 101.

references: "The business partner arranged a visa for a permanent stay, contacts, a warehouse close to the harbour and to the Roque Santeiro Market, the two most important points".[91]

Pepetela's book reveals the nepotism, corruption, illegal smuggling of diamonds, counterfeiting and general uncertainties of living in Luanda and the author writes with an ironical and sympathizing undertone, as a sharp observer of Angola's social and cultural reality. Despite the dramatic changes of the political system and economic development, this reality still refers to the continuities with the past, for which Jaime Bunda's family descent functions as the paradigmatical link. Slavery remembers the central position of the harbor in legal and illegal systems of global exchange.

9. Final observations

We have learned from the readings of the selected narrative texts that, on both sides of the ocean, the memory of slavery is a contemporary literary issue. The subjects of these historical processes—families of former slaves, former slaves themselves, or former slave-owning families—go into business and become decorators, entrepreneurs, writers, scientists, or detectives. For them, the lessons learnt in Salvador and Luanda with slavery have been crucial. Consequently, the urban space is constructed in accordance with the necessity of giving those roots a place, perceiving them as the "city within us," as the non-synchronic time-interspace of *oceanic modernity* in ongoing social relations and cultural practices.

The historical novel by Ana Maria Gonçalves, *Um defeito de cor*, is the only literary work in Brazil, which highlights Bahia's relationship with this "oceanic" dimension. Her female character, first victimized as a slave in Brazil, becomes a role model for urban modernity after having returned to the city of Ouidah in Benin and then of Lagos in Nigeria. Today, in these cities, the so-called Brazilian Quarter has been declared part of the National Heritage and the relation with Salvador da Bahia is recognized explicitly.[92]

91 "O sócio arranjou visto de permanência, contatos, um armazém não longe do porto nem do mercado Roque Santeiro, os dois pontos mais importantes." Ibid., p. 163.

92 Marjorie Moji Dolapo Alonge: *Afro-Brazilian Achitecture in Lagos State. A Case for Conservation*. Newcastle: Newcastle University, Doctoral Dissertation 1994; Marine de Souza: *Regard sur Ouidah. A Bit of History*. Ouidah: Souza 2000.

It became also clear that this role model character always finds stability in her spiritual orientation toward the water pantheon of African descent, even though living either in a Catholic or in a Protestant environment, throughout her long path of upward mobility. Gonçalves is certainly inspired by the important function of the *candomblé* and *terreiros* in Salvador today. Obviously, in Brazil, after many decades of continuous efforts to bring the cultural practices of African origin to public attention, they are no longer a taboo. They even claim to be considered an integral part of the Brazilian cultural heritage and the four-volume *Literatura e Afrodescendência no Brazil*, edited by Eduardo de Assis Duarte, is a breakthrough in this respect. Duarte offers numerous readings of Brazilian authors in past and present[93] and mentions that Gonçalves' female and black family saga is practically unheard of in Brazilian literature. A large publishing house accepted the book, and it even won the prestigious Casa de las Américas Prize for Brazilian literature in Havana in 2008, on which occasion the jury emphasized the high literary quality. Gonçalves' novel might even serve as an introduction to Brazil's various religious beliefs. The author repeatedly inserts many African expressions with an explanation, and in this way words like *ibeji, nkisi, agogôs* or *ewi*, among many others, become familiar to the readers. Our interpretation of *Um defeito de cor* demonstrates the influence of other writers such as Jorge Amado, António Olinto, José Saramago and Carolina Maria de Jesus. Selected scenes of the whale fishing station on the island Itaparica seem to be inspired by João Ubaldo Ribeiro's *Viva o povo brasileiro*, whose novel starts with the Battle of Pirajá on November 8, 1822 and the visit of Emperor Dom Pedro to Cachoeira in 1826, introducing multiple different characters. One of them is Zé Popo, a black Brazilian soldier of the Triple Alliance (Brazil, Uruguay, Argentina) against Paraguay in the Battle of Tutuita in May 1866. During the fight, Zé identifies with Oxossi, Ogum and Xango, the Yoruba gods.[94] However, Gonçalves' *Um defeito de cor* offers a female alternative to Ribeiro's male Yoruba pantheon; although he equally evokes the spiritual bond between the continents, he does not consider its fluid bindings.

93 Eduardo de Assis Duarte (ed.): *Literatura e Afrodescendência no Brazil: antologia crítica*, vol. 1: Precursores; vol. 2: Consolidação; vol. 3: Contemporaneidade; vol. 4: História, teoria, polémica, with Maria Nazareth Soares Fonseca. Belo Horizonte: AFMG 2011.

94 Ribeiro: *Viva o povo brasileiro*, pp. 437–454.

In her essay "Sea Foam – Toward a History of the Sea",[95] the art historian Hannah Baader calls attention to a thalassology, according to which art and science consider the sea as a thing that confers epistemic knowledge since the middle of the nineteenth century. Baader ascribes this interest to the advancement of technology in scientific research and visual art. Nevertheless, when applying this argument in *Um defeito de cor* to the historical meaning of the sea in such urban port societies as Salvador da Bahia, Ouidah or Lagos, Gonçalves makes clear that the sea's spiritual connotations have become part of its epistemological knowledge such as formulated by the UNESCO Convention for Safeguarding the Intangible Cultural Heritage.[96] In the face of increasing globalization, this Convention includes intangible traditions—living culture preserved and transmitted through its continued social practice—in her programs with the aim to encourage "the survival and vitality of the world's living local, national, and regional cultural heritage" embedded in ongoing social relationships.[97]
The presence of this intangible cultural heritage in contemporary and historical cultural practices is extensively documented by art historians, ethnologists and photographers such as Pierre Verger, who have portrayed the objects and rituals related to the Yoruba water pantheon, not only in Bahia but also in other places in Latin America and Africa. Yemayá is important enough that Abdias do Nascimento, the outstanding protagonist of the deconstruction of the myth of racial democracy in Brazil and the founder of the TEN-Theater (Teatro Experimental do Negro), named his daughter after her, dedicating the poem "Evocation of the Rose" to her in his volume *Axês do sangue da esperança (orikis)* (energies of the blood of hope (praise poems), 1984).[98]

95 Hannah Baader: Gischt. Zu einer Geschichte des Meeres. In: Hannah Baader / Gerhard Wolf (eds): *Das Meer, der Tausch und die Grenzen der Repräsentation.* Zurich: Diaphanes 2010, pp. 15–40.

96 This *Convention for the Safeguarding of the Intangible Cultural Heritage* was unanimously voted for at the meeting of the UNESCO General Conference in 2003. See Richard Kurin: Safeguarding Intangible Cultural Heritage in the 2003 UNESCO Convention: A Critical Appraisal. In: *Museum International*, 56,1–2 (2004), pp. 66–77.

97 Richard Kurin: Museums and Intangible Heritage: Culture Dead or Alive? In: *ICOM NEWS* 4 (2004), pp. 7–9, here p. 7.

98 Abdias Nascimento: Evocação da rosa (Para Yemanjá – no seu décimo aniversário). In: Id.: *Axês do sangue da esperança (orikis).* Rio de Janeiro: Axiamé 1983, pp. 32–34.

In spite of the relative popularity of *Um defeito de cor* as well as the Casa de las Américas Prize for Brazilian literature, Gonçalves's book is still placed in the realm of the Movimento Negro literature in Brazil. In the case of Luanda, meanwhile, this is different. Well-known and overall respected authors who are still very productive - such as Pepetela, Manuel Rui and Arnaldo Santos - identify with slave work. Thalassology here is here identified with Kianda, the name which is omnipresent in the urban landscape of Angola's capital.[99] Kianda appears in these narratives, obviously, in the effort of overcoming the "blocked memory" of the sea, according to Carmen Secco due to Portuguese colonization. Yet, we have seen that not only Portugal has to be addressed. In *A Gloriosa Família*, Pepetela demonstrates that, from the beginnings of its existence as a "lettered city" in the seventeenth century, Luanda has been related to Brazil and to other European nations.[100]

The deconstruction of this memory goes together with the emphasis on slavery in different historical periods. Pepetela's main character is clearly in disagreement with his situation but without possibilities to change his position in the rigid hierarchical construction of his slave work society. This rigid separation is dissolved in Arnaldo Santos' novel, drafted in a time that slavery is recently abolished. However, the citizens of Luanda still speak of "free" people, in other words, of people whose families were slaves. The *filhos-do-país* build up a network of local publications, protesting against the fact that, only three years after the official abolition of slavery in 1875, a "new slavery" of "forced labor" was installed, envisioning a law for bringing the "pretos livres" as contract workers—in other terms forced labor workers—to the plantations of São Tomé.[101] In his turn, José Eduardo Agualusa conceives his

99 Ineke Phaf-Rheinberger: Vom Meeresschaum und anderen Geheimnissen in der heutigen Literatur Angolas. In: Christoph Müller / Alexandre Martins / Lara Brück-Pamplona / Anne Begenat-Neuschäfer / Claudius Armbruster (eds): *Lusophone Konfigurationen*. Frankfurt am Main: Biblioteca Luso-Brasileira 2012, pp. 72–82.

100 Ineke Phaf-Rheinberger: A Rainha Njinga e os Holandeses na África Ocidental: Uma panorámica a partir da obra fundamental de Klaus Ratelband. In: *Mulemba. Revista Angolana de Ciências Sociais*, 3,5 (May 2013), pp. 161–171.

101 "Abolida que fora em 1875, a condição servil dos libertos, a sua sujeição ao trabalho contratado, a partir de 1878, tivera consequências perversas para os pretos livres, e tanto ele, como o chefe de Massangano sabiam de que leis tinham em mente." Santos: *A Casa Velha*, p. 176. Transl.: "Abolished in 1875 (slavery), the servile condition of the „free" people, their subjection to contract labor, since 1878, had perverse

novel *Nação Crioula* as Bhaba's "navigational allegory"[102] par excellence of this same time period and proclaims a radical renewal. He even lets Ana Olímpia travel from Brazil to Paris in 1877, where she makes a plea against slavery for an enlightened abolitionist audience.[103] Both authors suggest that, after the official abolition of slavery in Angola, this system remained virulent in the interstitial space between legal and illegal situations. Only at the end of the nineteenth century, Agualusa's former slave, the *returnee* Ana Olímpia, like Kehinde in *Um defeito de cor*, becomes a successful businesswoman in Luanda and then, equally, against the end of her life, writes her biography.

In his *Memória de mar* on contemporary Luanda, Manuel Rui points to the "degré zero" of writing after a significant historical change. The eponymous essay by Roland Barthes from 1953 discusses how the tradition of signifiers bound to the former situation makes it difficult to give shape to such a new perception.[104] For this reason, Rui chooses a scientific approach symbolized by the three expeditions and identifies the definitive liberation of slavery with independence. And in Nkeita's and Pepetela's novels on the period after 1991, when democracy was installed in the political system, we find a reflection on contemporary prosperous groups in Luanda who, nevertheless, are connected with the slavery past. They are now confronted with visitors and migrants to Angola from everywhere, in the context of the "translational global exchange" of knowledge and of the commodities in the warehouses close to the harbour and nearby the Roque Santero Market.

Undeniably, contemporary literature on Luanda touches upon the memory of the South-South maritime zone. Nevertheless, the portrayal of Angola's *paysage moralisé* shows a remarkable difference from Brazil. In Gonçalves' *Um defeito de cor*, we find the strict distinction of the "deficiency of color", which determinates social status throughout all levels of society, and this distinction is less explicitly addressed in the literature on Luanda. Here, apparently, the color line does not

consequences for the free blacks, and he as well as the chief of Masssangano knew which laws they had in mind." These *filhos-do-país* or the „naturaes" of Angola published an anonymous virulent protest against this forced labour in Portugal in 1901, reprinted in *Voz de Angola clamando no deserto*. Lisbon: Edições 70 / UEA 1984.

102 Bhabha: *The Location of Culture*, p. 8.

103 In Brazil, slavery was only abolished in 1888.

104 Roland Barthes: *Le degré zero de l'écriture*. Paris: Editions Gonthier 1971.

play the same role, not even in Pepetela's family of the seventeenth century. In addition, and as a second contrast, the Angolan remembrance of slavery is clearly male; we did not find a female writer who discusses this issue in a novel.[105] And, then, as a third remarkable point, we reencounter Kalunga, mentioned in Agostinho Neto's and John Bella's texts, in a different way. In Angola, Kianda seems to have replaced Kalunga for the remembrance of slavery from the perspective of mobility. Kalunga remains related to the underworld of the dead, such as in the poem "Kalunga Luigi, só os náufragos sabem" by Adriano Botelho de Vasconcelos.[106] This is not the case in the Americas. Martin Lienhard rightly observes in his comparative study *O mar e o mato* that, in Cuba, Kalunga obtained a more positive connotation than in Angola.[107] Agualusa conforms this observation, when letting a member of the crew on the *Nação crioula* explain to Ana Olímpia that, in Brazil, after having crossed the ocean, Kalunga is the resurrection from death:

> On the ship with which we escaped from Angola, the *Nação Crioula*, I met an old man who said that he had been friend with my father. He reminded me that, in our language (and in almost all the other languages of West Africa), the sea has the same name as death: Calunga. For the majority of the slaves, therefore, that journey was a passage through death. The life they left behind in Africa was Life; what they encountered in America or in Brazil, was a Renaissance.[108]

Other authors, such as the poet Lande Onawale from Salvador da Bahia and the anthropologist Custódia Wolney from Brasilia, reinforce this argument.[109]

105 Slavery is not addressed in the novel *A muxiluanda* by Maria Celestina Fernandez. See note 65.

106 Adriano Botelho Vasconcelos: Kalunga Luigi, só os náufragos sabem. In: Id. (ed.): *Todos os Sonhos. Antologia da Poesia Moderna Angolana*. Luanda: UEA 2005, p. 68.

107 Martin Lienhard: Mambo: canto ritual e memória escrava em Cuba. In: *O Mar e o Mato. Histórias da Escravidão*. Luanda: Kilombelombe 2005, pp. 33–70.

108 "No navio em que fugimos de Angola, o *Nação Crioula*, conheci um velho que afirmava ter sido amigo de meu pai. Ele recordou-me que na nossa língua (e em quase todas as outras línguas da África Ocidental) o mar tem o mesmo nome que a morte: Calunga. Para a maior parte dos escravos, portanto, aquela jornada era uma passagem através da morte. A vida que deixavam em África, era a Vida; a que encontravam na América ou no Brasil, um renascimento." Agualusa: *Nação crioula*, p. 157.

109 Lande Onawale: *Kalunga. Poemas de um mar sem fim / Poems of an Infinite Sea*, trans. by Raquel Luciana de Souza. Salvador da Bahia: edição do autor 2011; Custódia Wolney: *Eu, Kalunga*. São Paulo, Livro Pronto 2006.

In short, it has become clear that in contemporary fiction from Brazil and Angola, the "oceanic" spiritual intertextuality in the urban space of port cities implies concentrating on slavery. Whereas in other African narratives, slavery is still remembered in "forms of 'alternative memory' and metaphorization such as tragic repetition, fear and gossip, and tropes of suffering, bondage, and important sexuality",[110] in these literary texts slave work seems to be embedded in a metaphor for change. This might be explained because of the current economic prosperity of Brazil and Angola. However, I would argue that such an explanation is too restrictive. What might be at stake here is a developing consciousness of democracy, which measures "progress" or development with the abandonment of conditions of slavery. Due to this procedure, this oceanic dimension in literature becomes the carrier of the "intangible cultural heritage" of African descent and does not relate slavery exclusively to the past. It examines its memory in view of the different aspects of the production of modern spaces in historical and contemporary urban communities, as these books suggest, and this phenomenon is explored further in my current book project, in which this problematic is brought into the foreground as a relevant item in contemporary literary works in the Portuguese and Spanish languages in Africa and Latin America.

110 Laura Murphy: *Metaphor and the Slave Trade in West-African Literature*. Athens: Ohio University Press 2012, p. 3.

Bibliography

Afolabi, Niyi: *Afro-Brazilians Cultural Production in a Racial Democracy*. Rochester: University of Rochester Press 2009.

Agualusa, José Eduardo: *Nação crioula. A correspondência secreta de Fradique Mendes*. Lisbon: Dom Quixote 1997.

—: *Creole*, trans. by Daniel Hahn. London: Arcadia Books 2002.

De Alencastro, Luiz Felipe: *O trato dos viventes. A formação do Brasil no Atlantico Sul. Séculos XVI e XVII*. São Paulo: Companhia das Letras 2000.

Alonge, Marjorie Moji Dolapo: *Afro-Brazilian Achitecture in Lagos State. A Case for Conservation*. Newcastle: Newcastle University, Doctoral Dissertation 1994.

Amado, Jorge: *Bahia de todos os Santos. Guia de ruas e mistérios*, illustrated by Carlos Bastos. Rio de Janeiro: Record 1977 [1944].

—: *Mar Morto*. São Paulo: Companhia das Letras 2008 [1936].

Amenda, Lars / Malte Fuhrmann (eds): *Hafenstädte: Mobilität, Migration, Globalisierung* (= Special Issue, *Comparativ. Zeitschrift für Globalgeschichte und vergleichende Gesellschaftsforschung* 2,17 (2007)).

De Andrade, Mário: *Macunaíma o herói sem nenhum caráter*. Santiago de Chile: Editorial Universidad de Chile 1997 [1926].

Baader, Hannah: Gischt. Zu einer Geschichte des Meeres. In: Hannah Baader / Gerhard Wolf (eds): *Das Meer, der Tausch und die Grenzen der Repräsentation*. Zürich: Diaphanes 2010, pp. 15–40.

Barthes, Roland: *Le degré zero de l'écriture*. Paris: Éditions Gonthier 1971.

Bastin, Marie-Louise: *Arte Decorativa Cokwe*. Coimbra: Muséu Antropológico da Universidade de Coimbra, 2010.

Berman, Marshall: *All that is Solid Melts into Air. The Experience of Modernity*. London: Verso 1983.

Bhabha, Homi K. (ed.): *Nation and Narration*. London: Routledge 1990.

—: *The Location of Culture*. London: Routledge 1994.

Campos, Maria Consuelo Cunha: Luiz Gama. In: Eduardo de Assis Duarte (ed.): *Literatura e afrodescendencia no Brasil: antologia crítica*, vol. 1: Precursores. Belo Horizonte: Editora UFMG 2011, pp. 127–142.

Cartier, Carolyn: Cosmopolitism and the Maritime World City. In: *The Geographical Review* 89,2 (April 1999), pp. 278–289.

Caymmi, Stella: A parceria musical de Dorival Caymmi e Jorge Amado na adaptação teatral de *Terras do sem fim*. In: *Colóquio Jorge Amado, 70 anos de Mar morto*. Salvador da Bahia: Fundação Jorge Amado 2008, pp. 77–94.

Chaves, Rita / Carmen Secco / Tania Macêdo (eds): *Brasil / África. Como se o mar fosse mentira.* São Paulo: Editora UNESP 2006.

Coelho, Virgílio: Imagens, símbolos e representações, "Quiandas, Quitutas, Sereias": Imaginários locais, identidades regionais e alteridades. Reflexões sobre o quotidiano urbano luandense na publicidade e no universo do marketing. In: Id: *Os Túmúndòngò, os "Génios» da Natureza e o Kilàmbà. Estudos sobre a sociedade e a cultura Kímbùndù"*. Luanda: Kilombelombe 2010, pp. 142–194.

Couto, Mia: O mar que Jorge Amado criou não é o que banha no sentido geográfico, é mar que está dentro de nós, o mar que está vivo dentro de nossas vidas. http://www.expressodasilhas.sapo.cv/pt/noticias/go/-mia-couto (accessed October 14, 2013).

Duarte, Eduardo de Assis (ed.): *Literatura e Afrodescendência no Brazil: antologia crítica*, vol. 1: Precursores; vol. 2: Consolidação; vol. 3: Contemporaneidade; vol. 4: História, teoria, polémica, with Maria Nazareth Soares Fonseca. Belo Horizonte: AFMG 2011.

Evaristo, Conceição. In: Eduardo de Assis Duarte / Maria Nazareth Soares Fonseca (eds): *Literatura e Afrodescendência no Brasil: Antologia crítica*, vol. 4: História, teoria, polêmica. Belo Horizonte: Editora UFMG 2011, p. 105.

Fernandez, Maria Celestina: *A muxiluanda*. Luanda: Chá de Caxinde, 2008.

Ferreira, Manuel (ed.): *No reino de Calibán. Antologia panorâmica da poesia africana de expressão portuguesa*, vol. II: Angola, São Tomé e Príncipe. Lisbon: Seara Nova 1976.

Fundação Casa de Jorge Amado (ed.): *Colóquio Jorge Amado. 70 anos de Mar morto*. Salvador da Bahia: Fundação Jorge Amado 2008.

Gabaccia, Donna R. / Dirk Hoerder (eds): *Connecting Seas and Connected Ocean Rims. Indian, Atlantic, and Pacific Oceans and China Seas Migrations from the 1830s to the 1930s.* Leiden: Brill 2011.

Gilroy, Paul: *The Black Atlantic: Modernity and Double Consciousness.* Cambridge, MA: Harvard University Press 1995.

Gonçalves, Ana Maria: *Um defeito de cor. Romance.* Rio de Janeiro: Editora Record 2011 [2006].

Guri, Chó do: *Na boca árida da kyanda.* Luanda: Kilombelombe 2007.

De Jesus, Carolina Maria: *Quarto de despejo.* 8th ed. São Paulo: "Livr. Francisco Alves" / Paulo de Azevedo 1960.

—: *Child of the Dark: the Diary of Carolina Maria de Jesus*, trans. by David St. Clair. New York: New American Library 2003.

Júnior, Benjamin Abdala: Images of a Colonial Kalunga. In: *Research in African Literatures* 36,1 (2007), pp. 68–74.

Kurin, Richard: Safeguarding Intangible Cultural Heritage in the 2003 UNESCO Convention. A Critical Appraisal. In: *Museum International*, 56,1–2 (2004), pp. 66–77.

—: Museums and Intangible Heritage: Culture Dead or Alive? In: *ICOM NEWS* 4 (2004), pp. 7–9.

Lefebvre, Henri: *Introduction to Modernity: Twelve Preludes*. September 1959–1961, trans. by J. Moore. London: Verso 1995 [French, 1962].

—: *Writing on Cities*, ed. and trans. by E. Kofman / E. Lebas. Oxford: Basil Blackwell 1996.

Lienhard, Martin: Mambo: canto ritual e memória escrava em Cuba. In: *O Mar e o Mato. Histórias da Escravidão*. Luanda: Kilombelombe 2005, p. 33–70.

Macêdo, Tania: *Luanda, cidade e literatura*. São Paulo: Editora UNESP 2008.

De Macedo, Tânia Celestino: Visões do mar na literatura angolana contemporânea. http://www.fflch.usp.br/dlcv (accessed October 14, 2013).

Machado, Ana Maria: Pósfacio. A invenção da Bahia. In: Jorge Amado: *Mar Morto*. São Paulo: Companhia das Letras 2008, pp. 275–280.

Manzano, Juan Francisco: *Autobiografía del esclavo poeta y otros escritos*, ed., intr. and notes by William Luis. Madrid, Frankfurt am Main: Iberoamericana-Vervuert 2007.

Meisner, Christine: The Recovery of an Image: A Video Tale. In: Ineke Phaf-Rheinberger / Tiago de Oliveira Pinto (eds): *AfricAmericas. Itineraries, Dialogues, and Sounds*. Madrid / Frankfurt am Main: Iberoamericana-Vervuert 2008, pp. 69–76.

De Morães, Ana Cristina Moura Alves: O Mar sob o Ponto de Visto de Agostinho Neto no Conto *Náusea*. http://www.ueangola.com (accessed October 14, 2013).

Moura, Mariluce: Luiz Felipe de Alencastro: Brazil's Observer in the South Atlantic. http://revistapesquisa.fapesp.br/en/2011/10/01/luiz-felipe-de-alencastro-brazils-observer-in-the-south-atlantic/(accessed October 14, 2013).

Murphy, Laura: *Metaphor and the Slave Trade in West-African Literature*. Athens: Ohio University Press 2012.

Nascimento, Abdias: Evocação da rosa (Para Yemanjá – no seu décimo aniversário). In: Id.: *Axês do sangue da esperança (orikis)*. Rio de Janeiro: Axiamé 1983, pp. 32–34.

Neto, Agostinho: Kalumba. In: Id.: *Sagrada Esperança. Renuncia Impossível, Amanhecer*. Luanda: UEA 2009, p. 150.

Nkeita, Tazuary: *O último segredo*. 2nd ed. Luanda: UEA 2012.

Olinto, Antônio: *A Casa da Água*. São Paulo: Círculo do Livro 1975.

—: *Water House*, trans. by Dorothy Heapy. New York, NY: Carroll & Graf Publishers 1985.

OMIDIRE, Félix Ayoh': La identidad frente al poder. La asimetría de Yemayá en África y América Latina. In: Ineke Phaf-Rheinberger (ed.): *Historias enredadas. Representaciones asimétricas con vista al Atlántico*. Berlin: Ed. Tranvía 2011, pp. 145-165.

Onawale, Lande: *Kalunga. Poemas de um mar sem fim / Poems of an Infinite Sea*, trans. by Raquel Luciana de Souza. Salvador da Bahia: Edição do autor 2011.

Parreira, Adriano: *Dicionário de etnologia angolana*. Porto: Plural Editores 2013.

Pepetela: *O desejo de Kianda*. Lisbon: Dom Quijote 1995.

—: *A Gloriosa Família*. Lisbon: Dom Quixote 1997.

—: *Jaime Bunda, agente secreto. Estórias de alguns mistérios*. Rio de Janeiro, São Paulo: Record 2003.

Phaf-Rheinberger, Ineke: *The Air of Liberty. Narratives of the South Atlantic Past*. Amsterdam / New York: Rodopi 2008.

—: Vom Meeresschaum und anderen Geheimnissen in der heutigen Literatur Angolas. In: Christoph Müller Christoph Müller / Alexandre Martins / Lara Brück-Pamplona / Anne Begenat-Neuschäfer / Claudius Armbruster (eds): *Lusophone Konfigurationen*. Frankfurt am Main: Biblioteca Luso-Brasileira 2012, pp. 72–82.

—: A Rainha Njinga e os Holandeses na África Ocidental: Uma panorámica a partir da obra fundamental de Klaus Ratelband. In: *Mulemba. Revista Angolana de Ciências Sociais* 3,5 (May 2013), pp. 161–171.

Pólvora, Hélio: Jorge Amado e o romance do mar. In: *Colóquio Jorge Amado. 70 anos de Mar morto*. Salvador da Bahia: Fundação Jorge Amado 2008, pp. 53–60.

Reis, João José: *Rebelião escrava no Brasil. A história do levante dos Malês em 1835*, ed. revisada e ampliada. São Paulo: Companhia das Letras 2003.

Ribeiro, João Ubaldo: *Viva o povo brasileiro*. Rio de Janeiro: Nova Fronteira 1984.

Rimbaud, Arthur: „Adieu". In: Id.: *Sämtliche Dichtungen. Französisch und Deutsch*, ed. and trans. by Walther Küchler. Heidelberg: Schneider 1982., pp. 322–325.

Rui, Manuel: *Memória de mar*. Lisbon: Edições 70 1980.

Santos, Arnaldo: *A Casa Velha das margens*. 2nd ed. Luanda: Edições Chá de Caxinde 2010 [1999].

Schippers, Mineke: *Never Marry a Woman with Big Feet. Proverbs from Around the World*. New Haven / London: Yale University Press 2004.

Secco, Carmen Lúcia Tindó Ribeiro (ed.): *Antologia do Mar na Poesia Africana de Língua Portuguesa do Século XX. Angola*, vol. I. Luanda: Kilombelombe 2000.

Silva, Alberto da Costa: *Um rio chamado Atlântico. A África no Brasil e o Brasil na África*. Rio de Janeiro: Edições Nova Fronteira / UFRJ 2003.

Strickrodt, Silke: The Brazilian Diaspora to West Africa in the Nineteenth Century. In: Ineke Phaf-Rheinberger / Tiago de Oliveira Pinto (eds): *AfricAmericas. Itineraries, Dialogues, and Sounds*. Madrid / Frankfurt am Main: Iberoamericana-Vervuert 2008, pp. 36–68.

De Souza, Marine: *Regard sur Ouidah. A Bit of History*. Ouidah: Souza 2000.

Troni, Alfredo: *Nga Muturi. Cenas de Luanda*, preface by M. António. Lisbon: Edições 70 1973.

Vasconcelos, Adriano Botelho: Kalunga Luigi, só os náufragos sabem. In: Id. (ed.): *Todos os Sonhos. Antologia da Poesia Moderna Angolana*. Luanda: UEA 2005, p. 68.

Voz de Angola clamando no deserto. Lisbon: Edições 70 / UEA 1984. Reprint of the 1901 edition.

Wolney, Custódia: *Eu, Kalunga*. São Paulo: Livro Pronto 2006.

TYPVS ORB
AMERICA SIVE IN DIA NOVA. Ao 1492. a Christophoro Colombo nomine regis Castellæ primum detecta.
CIRCVLVS ARCTICVS
ANIAN regnum
QVIVIRA regnu
Tolm
Totonteac
Marata
Cicuicas
Tagil
Chilaga
Noua Francia
Florida
Estotilant
TROPICVS CANCRI
MAR DEL NORT
Hispania noua
CIRCVLVS AEQVINOCTIALIS
Caribana
Peru
Brasil
Amazones
Tisnada
MAR DEL ZVR
EL MAR PACIFICO
TROPICVS CAPRICORNI
Chica
Chile
Rio de la Plata
Terra del Fuego
CIRCVLVS ANTARCTICVS
Hanc continentem Australem, nonnulli Magellanicam regionem ab eius inuentore nuncupant.
Noua Guinea nuper inuenta quæ an sit insula an pars continentis Australis incertu est
Archipelago di
Las dos hermanos
Los Bolcanes
La farfana
Malabrigo
Lazaro
Abreojo
Rocca partida
Ins di los Tiburones
I. de los pescadores
190 200 210 220 230 240 250 260 270 280 290 300 310 320 330
TERRA AVSTR
QVID EI POTEST VIDERI MAGNVM IN
OMNIS, TOTIVSQVE MVNDI N